VISIT US AT

www.syngress.co

Syngress is committed to publishing high-quality bo
delivering those books in media and formats that fit
tomers. We are also committed to extending the util ... purchase
via additional materials available from our Web site.

SOLUTIONS WEB SITE

To register your book, visit www.syngress.com/solutions. Once registered, you can access our solutions@syngress.com Web pages. There you may find an assortment of value-added features such as free e-books related to the topic of this book, URLs of related Web sites, FAQs from the book, corrections, and any updates from the author(s).

ULTIMATE CDs

Our Ultimate CD product line offers our readers budget-conscious compilations of some of our best-selling backlist titles in Adobe PDF form. These CDs are the perfect way to extend your reference library on key topics pertaining to your area of expertise, including Cisco Engineering, Microsoft Windows System Administration, CyberCrime Investigation, Open Source Security, and Firewall Configuration, to name a few.

DOWNLOADABLE E-BOOKS

For readers who can't wait for hard copy, we offer most of our titles in downloadable Adobe PDF form. These e-books are often available weeks before hard copies, and are priced affordably.

SYNGRESS OUTLET

Our outlet store at syngress.com features overstocked, out-of-print, or slightly hurt books at significant savings.

SITE LICENSING

Syngress has a well-established program for site licensing our e-books onto servers in corporations, educational institutions, and large organizations. Contact us at sales@syngress.com for more information.

CUSTOM PUBLISHING

Many organizations welcome the ability to combine parts of multiple Syngress books, as well as their own content, into a single volume for their own internal use. Contact us at sales@syngress.com for more information.

SYNGRESS®

STUDY GUIDE AND PRACTICE EXAM

Dr. Paul Sanghera

KEY	SERIAL NUMBER
001	HJIRTCV764
002	PO9873D5FG
003	829KM8NJH2
004	9516L2JHT2
005	CVPLQ6WQ23
006	VBP965T5T5
007	HJJJ863WD3E
008	2987GVTWMK
009	629MP5SDJT
010	IMWQ295T6T

PUBLISHED BY
Syngress Publishing, Inc.
800 Hingham Street
Rockland, MA 02370

RFID+ Study Guide and Practice Exam

1 2 3 4 5 6 7 8 9 0
ISBN-10: 1-59749-134-9
ISBN-13: 978-1-59749-134-1

Publisher: Andrew Williams
Acquisitions Editor: Erin Heffernan
Technical Editor: Francesco Kung Man Fung
Cover Designer: Michael Kavish
Page Layout and Art: Patricia Lupien
Copy Editor: Darlene Bordwell
Indexer: Richard Carlson

Distributed by O'Reilly Media, Inc. in the United States and Canada.
For information on rights, translations, and bulk sales, contact Matt Pedersen, Director of Sales and Rights, at Syngress Publishing; email matt@syngress.com or fax to 781-681-3585.

Printed in Canada

Acknowledgments

Syngress would like to acknowledge the following people for their kindness and support in making this book possible.

Syngress books are now distributed in the United States and Canada by O'Reilly Media, Inc. The enthusiasm and work ethic at O'Reilly are incredible, and we would like to thank everyone there for their time and efforts to bring Syngress books to market: Tim O'Reilly, Laura Baldwin, Mark Brokering, Mike Leonard, Donna Selenko, Bonnie Sheehan, Cindy Davis, Grant Kikkert, Opol Matsutaro, Mark Wilson, Rick Brown, Tim Hinton, Kyle Hart, Sara Winge, Peter Pardo, Leslie Crandell, Regina Aggio Wilkinson, Pascal Honscher, Preston Paull, Susan Thompson, Bruce Stewart, Laura Schmier, Sue Willing, Mark Jacobsen, Betsy Waliszewski, Kathryn Barrett, John Chodacki, Rob Bullington, Kerry Beck, Karen Montgomery, and Patrick Dirden.

The incredibly hardworking team at Elsevier Science, including Jonathan Bunkell, Ian Seager, Duncan Enright, David Burton, Rosanna Ramacciotti, Robert Fairbrother, Miguel Sanchez, Klaus Beran, Emma Wyatt, Krista Leppiko, Marcel Koppes, Judy Chappell, Radek Janousek, Rosie Moss, David Lockley, Nicola Haden, Bill Kennedy, Martina Morris, Kai Wuerfl-Davidek, Christiane Leipersberger, Yvonne Grueneklee, Nadia Balavoine, and Chris Reinders for making certain that our vision remains worldwide in scope.

David Buckland, Marie Chieng, Lucy Chong, Leslie Lim, Audrey Gan, Pang Ai Hua, Joseph Chan, June Lim, and Siti Zuraidah Ahmad of Pansing Distributors for the enthusiasm with which they receive our books.

David Scott, Tricia Wilden, Marilla Burgess, Annette Scott, Andrew Swaffer, Stephen O'Donoghue, Bec Lowe, Mark Langley, and Anyo Geddes of Woodslane for distributing our books throughout Australia, New Zealand, Papua New Guinea, Fiji, Tonga, Solomon Islands, and the Cook Islands.

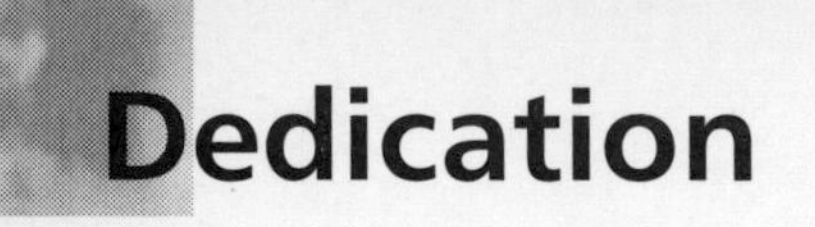

Dedication

To my brother Surinder's wisdom
Uncorrupted by academic degrees

Author

Paul Sanghera, an expert in multiple fields including computer networks and physics (the parent fields of RFID), is a subject matter expert in RFID. With a Masters degree in Computer Science from Cornell University and a Ph.D. in Physics from Carleton University, he has authored and co-authored more than 100 technical papers published in well reputed European and American research journals. He has earned several industry certifications including RFID+, Network+, Linux+, PMP, CAPM, Project+, SCBCD, and SCJP. Dr. Sanghera has contributed to building world-class technologies such as Netscape Communicator and Novell's NDS. He has taught technology courses at various institutes including San Jose Sate University and Brooks College. As an engineering manager, he has been at the ground floor of several startups. The best selling author of several books in technology and project management, Dr. Sanghera is currently the President of Infonential, Inc, an information products and services company specializing in project management and emerging technologies such as RFID and nanotechnology. For more information on Dr. Sanghera, or to contact him, you can visit the website www.infonentialinc.com.

Technical Editor

Francesco Kung Man Fung (SCJP, SCWCD, SCBCD, ICED, MCP, OCP) has worked with Java, C#, and ASP.net for 6 years. Mainly, he develops Java-based/.net financial applications. He loves to read technical books and has reviewed several certification books.

Fung received a Bachelor's and a Master Degree in Computer Science from the University of Hong Kong.

Author's Acknowledgments

As they say (well, if they don't any more, they should), first thing first. Let me begin by thanking David Fugate and Andrew Williams who triggered this project. With two thumbs up, thanks to Erin Heffernan, the project manager of this book project, for her focus, dedication, professionalism, and results-oriented approach.

It takes a team to materialize a book idea into a published book. It is my great pleasure to acknowledge the hard and smart work of the Syngress team that made it happen. Here are a few names to mention: Darlene Bordwell for copy editing, Patricia Lupien for page layout and art, and Richard Carlson for Indexing. I am thankful to Francesco Kung, the technical editor of this book, for doing an excellent job in thoroughly reviewing the manuscript and offering valuable feedback. Also I'm thankful to Corey Cotton for useful comments and suggestions.

In some ways, writing this book is an expression of the technologist and educator inside me. I thank my fellow technologists who guided me at various places during my journey in the computer industry from Novell to Dream Logic: Chuck Castleton at Novell, Delon Dotson at Netscape and MP3.com, Kate Peterson at Weborder, and Dr. John Serri at Dream Logic. I also thank my colleagues and seniors in the field of education for helping me in so many ways to become a better educator. Here are a few to mention: Dr. Gerald Pauler (Brooks College), Professor David Hayes (San Jose State University), Professor Michael Burke (San Jose State University), and Dr. John Serri (University of Phoenix).

Friends always lend a helping hand, in many visible and invisible ways, in almost anything important we do in our lives. Without them, the world would be a very boring and uncreative place. Here are a few I would like to mention: Stanley Wong, Patrick Smith, Kulwinder, Major Bhupinder Singh Daler, Ruth Gordon, Srilatha Moturi, Baldev Khullar, and the Kandola family (Gurmail and Sukhwinder).

Last, but not least, my appreciation (along with my heart) goes to my wife, Renee, and my son, Adam, for not only peacefully coexisting with my book projects but also supporting them.

Contents

Foreword

Introduction

> How would you like it if, for instance, one day you realized your underwear was reporting on your whereabouts?
>
> — Debra Bowen, California State Senator, at a 2003 hearing

In this book, you will not only learn the basics of radio frequency identification (RFID), but also prepare for the CompTIA RFID+ certification exam in the process of doing so: two in one. In other words, this book covers the topics determined by the exam objectives for the CompTIA RFID+ certification exam, RF0-001. Each chapter explores topics in RFID specified by a set of exam objectives in a manner that makes the presentation cohesive, concise, and yet comprehensive.

Who This Book is For

This book is primarily targeted at the RFID professionals and students who want to prepare for the CompTIA RFID+ exam, RF0-001. Since the book has a laser-sharp focus on the exam objectives, expert RFID professionals who want to pass the exam can use this book to ensure that they do not overlook any objective. Yet, it is not an exam-cram book. The chapters and the sections inside each chapter are presented in a logical learning sequence: Every new chapter builds upon knowledge acquired in previous chapters, and there is no hopping from topic to topic. The concepts and topics, simple and complex, are explained in a concise yet comprehensive fashion. This facilitates stepwise learning and prevents confusion. Furthermore, chapter 1 presents very basic introduction to physics and math concepts relevant to learning RFID for the absolute beginners. Hence, this book is also very useful for beginners to get up to speed quickly even if they are new to RFID and do no have the necessary physics and math background. Even after the exam, you will find yourself returning to this book as a useful and practical reference for basics of RFID.

In a nutshell, this book can be used by the following audiences:

- RFID professionals and students who want to prepare for the CompTIA RFID+ exam
- RFID professionals who are looking for a quick and practical RFID reference
- Beginners who want to join the RFID profession
- Instructors who want to offer a basic course on RFID

How this Book is Structured

The structure of this book is determined by the following two requirements:

- The book is equally useful for both the beginners and the experts who want to pass the CompTIA RFID+ exam.
- Although it has a laser sharp focus on the exam objectives, the book is not an exam cram. It presents the material in a logical learning sequence so that the book can be used for learning (or teaching) basics of RFID.

With the exception of the introductory chapter 1, each chapter begins with a list and explanation of exam objectives on which the chapter is focused. We have somewhat rearranged the order of the exam domains to keep the topics and the subject matter in line with sequential learning and to avoid hopping from topic to topic.

The first section in each chapter is the Introduction, in which we establish the concepts or topics that will be explored in the chapter. As you read through a chapter, you will find the following features:

- **Note.** Notes emphasize important concepts or information
- **Caution**. Cautions point out information that may be contrary to your expectations depending upon your level of experience with the

Java programming. Both Notes and Alerts are important from the exam viewpoint.

- **Tip**. Provides additional real-world insight into the topic being discussed.
- **Exercise**. Exercises are designed to help you understand how some concepts work.
- **Key Terms**. This section lists the important terms and concepts introduced in the chapter along with their definitions.
- **Summary.** This section provides the big picture and reviews the important concepts in the chapter.
- **Exam's-Eye View.** This section highlights the important points in the chapter from the perspective of the exam: the information that you must comprehend, the things that you should watch out for because they might not seem to go along with the ordinary order of things, and the facts that you should memorize for the exam.
- **Self Test.** has a two-pronged purpose: to help you test your knowledge about the material presented in the chapter and to help you evaluate your ability to answer the exam questions based on the exam objectives covered in the chapter. The answers to the Self Test questions are presented in Appendix B.

Other special features of the book are the following:

- A complete practice exam with questions modeled after the real exam and fully explained answers.
- Detailed answers to all the Self Test questions and exercises.
- A glossary that contains definitions of key RFID terms and concepts.

Prerequisites

Neither the CompTIA RFID+ exam nor this book has any pre-requisite.

About the RFID+ exam

Neither the physics behind it, nor the RFID technology itself is new. But it's only recently that the greatness has been bestowed upon RFID by the giant influencers such as U.S. Department of Defense and Wal-Mart in their mandates, and in a flurry of industrial mandates that followed. Now armed with these mandates, government legislations, and the resulting hyperbole, RFID has set its journey to change the world. With the market for RFID services projected to exceed $4 billion by 2008, a late start by a corporation in evaluating and implementing the technology could turn into a competitive disadvantage. Taking on the opportunity, the Computing Technology Industry Association (CompTIA) has launched the RFID+ certification in an effort to develop the workforce and provide the industry with a standard for measuring competency in the installation and maintenance of RFID.

Topics Covered in the RFID+ Exam

The topics covered in the exam and their relative weights are listed in the following table.

Proportion of Questions from Each Domain

Domain #	Domain Name	Percentage Coverage in the Exam	Approximate Number of Questions
1.0	Interrogation zone basics	13	10
2.0	Testing and troubleshooting	13	10
3.0	Standards and regulations	12	10
4.0	Tag knowledge	11	9
5.0	Design selection	11	9
6.0	Installation	11	9
7.0	Site analysis	11	9
8.00	RF Physics	11	9
9.0	RFID peripherals	7	6
	Total	**100**	**81**

Preparing for the RFID+ exam

According to CompTIA, the skills and knowledge measured by this examination are derived from an industry-wide job task analysis (JTA) and have been validated by Subject Matter Experts from around the globe. The CompTIA RFID+ certification proves that you have the foundational RFID knowledge, and a minimum of 6 to 24 months of experience in

RFID or a related industry with competencies including the following:

- Installation, configuration, and maintenance of RFID or related hardware and device software
- Site survey/site analysis
- RFID design selection

If you are a beginner, you will learn RFID while preparing for the exam because this book is not a mere exam cram. On the other end of the spectrum, even an RFID expert may fail this exam if not prepared for it properly. So, experts can use this to make sure they don't miss any exam objective. From the exam point of view, pay special attention to the following items while preparing for the exam:

1. Carefully read the exam objectives in the beginning of each chapter.
2. Make sure you understand the Notes, Cautions, and Exercises in each chapter.
3. Study the review questions at the end of each chapter.
4. Take the practice exam that comes with this book toward the end your exam preparation.
5. Review the Exam's-Eye View sections during the last hours of your preparation.

Taking the RFID+ exam

The RFID+ certification consists of one exam available at authorized Prometric Testing Centers throughout the world. Following are some important details of the exam:

- Exam ID: RF0-001
- Prerequisite: None
- Cost: $190 for CompTIA members, $237 fro non-members (The cost may vary by country and also if you have discount coupons.)

- Number of questions: 81
- Maximum time allowed: 90 minutes
- Minimum Pass score: 630 on the scale of 100-900

The question types are multiple choice including drag and drop. In most of the questions, you are asked to select the correct answers from multiple answers presented for a question. The number of correct answers is given.

For the current and complete information, you can visit the CompTIA site: www.comptia.org

Best wishes for the exam. Go for it!

Contacting the Author

More information about Dr. Paul Sanghera can be found at: www.paulsanghera.com

He can be reached at: paul_s_sanghera@yahoo.com

Exam Readiness Checklist

Exam Objective	Chapter #
Domain 1.0 Interrogation zone basics	4
1.1 Describe interrogator functionality	
1.1.1 I/O capability	
1.1.2 Hand-held interrogators	
1.1.3 Vehicle mount interrogator	
1.1.4 LAN/Serial communications	
1.1.5 Firmware upgrades	
1.1.6 Software operation (GUIs)	
1.2 Describe configuration of interrogation zones	
1.2.1 Explain interrogator to interrogator interference	
1.2.2 Optimization	
1.2.3 System performance and tuning	
1.2.4 Travel speed and direction	
1.2.5 Bi-static / monostatic antennas	

Exam Objective	Chapter #
1.3 Define anti-collision protocols (e.g., number of tags in the field/response time)	
1.4 Given a scenario, solve dense interrogator environment issues (domestic/international)	
1.4.1 Understand how a dense interrogator installation is going to affect network traffic	
1.4.2 Installation of multiple interrogators, (e.g., dock doors, synchronization of multiple interrogators, antenna footprints)	
Domain 2.0 Testing and Troubleshooting	**10**
2.1 Given a scenario, troubleshoot RF interrogation zones (e.g., root-cause analysis)	
2.1.1 Analyze less than required read rate	
2.1.1.1 Identify improperly tagged items	
2.1.2 Diagnose hardware	
2.1.2.1 Recognize need for firmware upgrades	
2.1.3 Equipment replacement procedures (e.g., antenna, cable, interrogator)	
2.2 Identify reasons for tag failure	
2.2.1 Failed tag management	
2.2.2 ESD issues	
2.3 Given a scenario, contrast actual tag data to expected tag data	
Domain 3.0 Standards and Regulations	**5**
3.1 Given a scenario, map user requirements to standards	
3.1.1 Regulations, standards that impact the design of a particular RFID solution	
3.2 Identify the differences between air interface protocols and tag data formats	
3.3 Recognize regulatory requirements globally and by region (keep at high level, not specific requirements — may use scenarios)	
3.4 Recognize safety regulations/issues regarding human exposure	

Exam Objective	Chapter #
Domain 4.0 Tag Knowledge	**3**
4.1 Classify tag types	
4.1.1 Select the RFID tag best suited for a specific use case.	
4.1.1.1 Pros and cons of tag types	
4.1.1.2 Tag performance	
4.1.1.2.1 Tag antenna to region/frequency	
4.1.2 Identify inductively coupled tags vs. back-scatter	
4.1.3 Identify the differences between active and passive	
4.2 Given a scenario, select the optimal locations for an RFID tag to be placed on an item.	
4.2.1 Evaluate media and adhesive selection for tags	
4.2.2 Tag orientation and location	
4.2.2.1 Tag stacking (shadowing)	
4.2.3 Package contents	
4.2.4 Packaging	
4.2.4.1 Items	
4.2.4.2 Tags	
4.2.4.3 Labels	
4.2.4.4 Inserts	
4.2.5 Liquids	
4.2.6 Metal	
4.2.7 Polarization	
Domain 5.0 Design Selection	**6**
5.1 Given a scenario, predict the performance of a given frequency and power (active/passive) as it relates to: read distance, write distance, tag response time, storage capacity	
5.2 Summarize how hardware selection affects performance (may use scenarios)	

Exam Objective	Chapter #
5.2.1 Antenna type	
5.2.2 Equipment mounting and protection	
5.2.3 Cable length/loss	
5.2.4 Interference considerations	
5.2.5 Tag type (e.g., active, passive, frequency)	
Domain 6.0 Installation	8
6.1 Given a scenario, describe hardware installation using industry standard practices	
6.1.1 Identify grounding considerations (e.g., lightning, ground loops, ESD)	
6.1.2 Test installed equipment and connections (pre-install and postinstall)	
6.2 Given a scenario, interpret a site diagram created by a RFID architect describing interrogation zone locations, cable drops, device mounting locations	
Domain 7.0 Site Analysis (i.e., before, during and after installation)	7
7.1 Given a scenario, demonstrate how to read blueprints (e.g., whole infrastructure)	
7.2 Determine sources of interference	
7.2.1 Use analysis equipment such as a spectrum analyzer, determine if there is any ambient noise in the frequency range that may conflict with the RFID system to be installed	
7.3 Given a scenario, analyze environmental conditions end-to-end	
Domain 8.0 RF Physics	2
8.1 Identify RF propagation/communication techniques	
8.2 Describe antenna field performance/characteristics as it relates to reflective and absorptive materials (may use scenarios)	
8.3 Given a scenario, calculate radiated power output from antenna based on antenna gains, cable type, cable length, interrogator transmit power (include formulas in scenario)	

Exam Objective	Chapter #
Domain 9.0 RFID Peripherals	**9**
9.1 Describe installation and configuration of RFID printer (may use scenarios)	
9.2 Describe ancillary devices/concepts	
9.2.1 RFID printer encoder	
9.2.2 Automated label applicator	
9.2.3 Feedback systems (e.g., lights, horns)	
9.2.4 RTLS	

Chapter 1

RFID+

Physics, Math, and RFID: Mind the Gap

Learning Objectives

- Understand basic physics concepts such as energy, force, field, power, speed, work, physical quantity, and units.
- Understand electricity, magnetism, and electromagnetism.
- Understand electromagnetic waves and the basic wave properties such as frequency and wavelength.
- Identify different kinds of electromagnetic waves in the electromagnetic spectrum.
- Learn how to perform some basic math operations: powers of 10, logarithms, and some unit conversions.
- Identify the difference between barcode technology and RFID.

Introduction

What do the U.S. Department of Defense, Wal-Mart, and you have in common? Radio frequency identification, or RFID! Whether you choose to know about it or not, RFID affects you and the world around you in a ubiquitous way. So, congratulations that you have chosen to learn about it.

The first thing to understand about RFID is that it is an application of physics to the extent that the core functioning of RFID technology is governed by the laws of physics. You don't need to have a Ph.D. in physics to become a successful RFID professional, but an understanding of the physics of RFID will enable you to design, deploy, and operate RFID systems in an optimal way. In this chapter, we attempt to ease your way into physics as it relates to RFID by explaining some basic physics concepts. As they say, mathematics is the language of physics, or of any science for that matter. The good news is that you need only very simple math to understand RFID: powers of 10, logarithms, and some unit conversions. Before you dive into the book, we take a bird's-eye view of RFID in this chapter. The goal is to provoke you to start asking questions about the details that will be addressed in the forthcoming chapters.

The overall goal of this chapter is to help you avoid falling into the gaps between physics, math, and RFID. We fill those gaps by exploring three avenues: basic physics concepts, the math of RFID, and an overview of RFID.

Some Bare-Bones Physics Concepts

Just when you thought you got away with missing physics classes in high school, here comes a physics lecture for you! But fear not. It's going to be very simple and concise.

As you already know, physics is a discipline in natural science. The word *science* has its origin in a Latin word that means *to know*. Science is the body of knowledge of the natural world, organized in a rational and verifiable way. The word *physics* has its origin in the Greek word that means *nature*. Physics is that branch (or discipline) of science that deals with understanding the universe and its systems in terms of fundamental constituents of matter (such as atoms, electrons, and quarks) and the interactions among those constituents. *Applied physics* refers to the practical (such as technological) use of physics—for example, electronics, engineering, and RFID. In other words, applied physics involves utilizing basic physics principles to build practical devices and systems such as radios, televisions, cellular phones, or an RFID system.

To clear your way toward understanding the physics behind RFID, let's look at some basic physics concepts:

- **Physical quantity** A measurable observable is called a *physical quantity*. In physics, we understand the universe and the systems in the universe in terms of physical quantities and the relationships among them. In other words, laws of physics are expressed in terms of relationships among the physical quantities.

Length, time, speed, force, energy, and temperature are some examples of physical quantities.

- **Unit** A physical quantity is measured in numbers of a basic amount called a *unit*. The measurement of a quantity contains a number and a unit—for example, in 15 miles, *mile* is a unit of distance (or length).
- **Force** This is the influence that an object exerts on another object to cause some change.
- **Interaction** This is a mutual force between two objects through which they affect each other. For example, two particles attract each other or repel each other. Sometimes the words *interaction* and *force* are used synonymously. There are four known basic interactions (or forces) that keep the universe functioning together:
    - Gravitational force
    - Electromagnetic force
    - Strong nuclear force
    - Weak nuclear force
    Where there is a force, there is energy, or potential for energy.
- **Energy** Energy is the measure of the ability of a force to do work. There are different kinds of energies corresponding to different forces, such as electromagnetic energy.
- **Power** Power is the amount of work done or the energy transferred per unit time.
- **Work** Work is a measure of the amount of change produced by a force acting on an object. But how is it possible that two objects separated from each other can exert force on each other? This is where the concept of field comes into the picture.
- **Field** The basic forces of nature work between two objects without the objects physically touching each other. For example, Sun and Earth attract each other through gravitation force without touching each other. This effect is called *action at a distance* and is explained in physics by the concept of a *field*. The two objects (which, for example, attract or repel each other from a distance) create a field in the space between them, and it is that field that exerts the force on the objects. For example, there is a gravitation field corresponding to gravitational force and an electromagnetic field corresponding to electromagnetic force.

- **Speed** Speed, in general, means the rate of something. In physics, it means the rate of motion; for example, your car is moving at a speed of 70 miles per hour.
- **Hypothesis** A hypothesis is a principle-like statement made as an explanation of a phenomenon and is generally based on previous observations, extensions of existing scientific theories, or both. The scientific method requires that a scientific hypothesis must be verifiable; that is, you must be able to test it. The word *hypothesis* has its roots in the Greek word that means *to suppose*.
- **Law** A physics law (also called a physical law, a law of nature, or a scientific law) is a set of generalized conclusions based on observations of physical behavior through repeated scientific experiments, and these conclusions are generally accepted within the scientific community. A hypothesis may turn into a law through repeated confirmation by scientific experiments.

Of the four basic interactions in the universe, the interaction that is relevant to RFID is the electromagentic interaction, which exhibits itself in our world in many forms, including electricity and magnetism.

Understanding Electricity

Electricity is the property of matter related to electric charge. Historically, the word *electricity* has been used by several scientists to mean electric charge. This property (electricity) is responsible for several natural phenomena such as lightning and is used in several industrial applications such as electric power and the whole field of electronics.

To understand electricity, you must understand the related concepts discussed in the following:

Electric charge Electric charge, also referred to simply as *charge*, is a basic property of some fundamental particles of matter. There are two types of charge: positive and negative. For example, an electron has a negative charge, and a positron (an anti-particle of electron) has a positive charge. The standard symbol used to represent charge is q or Q. Two particles (or objects) with the same type of charge repel each other, and two objects with the opposite types of charge attract each other. The charge is measured in units of *coulomb*, denoted by C.

Electric potential/voltage The electric potential difference between two points is the work required to take one unit, C, of charge from one point to another. This is commonly called *electric potential* or *voltage* because it's measured in units of *volt*, denoted by V.

Capacitance This is the amount of charge stored in a system, called a *capacitor*, per unit of electric potential. In other words, the capacitance, *C*, is defined by the following equation:

$$C = Q/V$$

One example of a capacitor is the so-called parallel plates capacitor: two metallic plates separated from each other, with each plate carrying equal and opposite charge, *Q*, with a potential difference between them, *V*. Capacitance is measured in units of *farad*, denoted by *F*. For example, if the charge on each plate of a parallel plate capacitor is one C, and the voltage between them is one V, the capacitance of the capacitor will be one F.

Electric current This is the rate of flow of electric charge per unit time and can be defined by the following equation:

$$I = Q/t$$

In this equation, *I* is the current and *Q* is the amount of charge that flowed past a point in time *t*. Current is measured in units of ampere, denoted by *A*. For example, one *C* of charge flowing past a point in one second represents one *A* of current. The material such as metals that permit relatively free flow of charge are called *conductors*, whereas the materials such as glass that do not allow free flow of charge are called *insulators*.

Resistance This is a measure of opposition offered by a material to the flow of charge through it. The resistance can be measured by the following equation:

$$I = V/R$$

This means the larger the resistance, the smaller the current. Resistance is measured in units of ohm, denoted by Ω. For example, if the voltage of one V creates one A of current in a conductor, then the resistance of the conductor is one Ω.

Electric energy This is the amount of work that can be done by an amount of electric charge across a potential difference. For example, the energy, *E*, of a charge *Q* across a voltage *V* is given by the following equation:

$$E = QV$$

Electric power This is the rate of work performed by an electric current. In other words, it's the electric energy produced or consumed per unit of time, and is given by the following equation:

$$P = E/t = QV/t = IV$$

The power is measured in units of watt (*W*). For example, the power consumed to maintain a current of one *A* across a voltage of one *V* is one *W*.

Exercise 1.1

Show that electric power can also be expressed by the following equations:

$$P = I^2R$$
$$P = V^2/R$$

Solution: We know that:

$P = IV$

We also know that:

$I = V/R$

Therefore:

$P = IV = (V/R)V = V^2R$

But:

$I = V/R$ means $V=IR$

Therefore:

$P = IV = I \times IR = I^2R$

Electric field Electric field is a field that charges at a distance used to exert force on each other. In other words, the charges at a distance interact with each other through their fields, called *electric fields*.

Two charges of the same type exert repulsive force on each other, and two charges of opposite types exert attractive force on each other, and this force is called *electric force*. A charge in motion creates another kind of force, called *magnetic force*.

Understanding Magnetism

Magnetism is the property of material that enables two objects to exert a specific kind of force on each other, called *magnetic force*, which is created by electric charge in motion. To understand magnetism, you must understand the related concepts discussed in the following:

Magnetic field A magnetic field is a field produced by a moving charge that it uses to exert magnetic force on another moving charge.

Magnetic flux This is a measure of the quantity of magnetic field through a certain area. It is proportional to the strength of the magnetic field and the surface area under consideration. For example, the current running through a wire in a circuit will create the magnetic field and hence the magnetic flux in the area around it.

Faraday's Law Faraday's Law states that the change in magnetic flux creates electromotive force, which is practically a voltage. In other words, the changing magnetic flux through a circuit will induce a current in the circuit. Recall that the magnetic flux can be created by the current in a circuit. Faraday's Law says the reverse: The change in flux can create current.

Inductive coupling Consider two electric circuits next to each other. There will be magnetic flux through the second circuit due to the current in the first circuit. If you change the current in the first circuit, it will change the magnetic flux through the second circuit, and the change in magnetic flux will create the current through the second circuit due to Faraday's Law. This effect, called *inductive coupling*, is used in RFID systems. You will see in this book that readers use inductive coupling to communicate with passive tags in an RFID system. You will be introduced to readers and tags later in this chapter.

Electricity and magnetism are related to each other and can be looked upon as two facets of what is called *electromagnetism*.

Understanding Electromagnetism

Electromagnetism is the unified framework through which to understand electricity, magnetism, and the relationship between them—in other words, to understand electric fields and magnetic fields and the relationship among them. To see the relationship, first recall that a charge creates an electric field and that when the same charge starts moving, it creates a magnetic field. The electric field exerts electric force, whereas a magnetic field exerts magnetic force; both originate from the electric charge. Therefore, they are intimately related: A changing electric field produces a magnetic field, and a changing magnetic field produces an electric field. Due to this intimacy, the electric force and magnetic force are considered two different manifestations of the same unified force, called *electromagnetic (EM) force*. The unified form of the electric field and magnetic field is called an *electromagnetic field*, and the electric field and the magnetic field are considered its components. In other words, electromagnetic force is exerted by an electromagnetic field.

Where there is a force, there is energy. The energy corresponding to electromagnetic force is called *electromagnetic energy* or *electromagnetic radiation*. This energy is transferred from one point in space to another point through what are called *electromagnetic waves*.

Electromagnetic Waves

A *wave* is a disturbance of some sort that propagates through space and transfers some kind of energy from one point to another. For example, when you speak to a person face to face, the sound wave travels from your mouth to the ear of the listener. The "disturbance" here is the change of pressure in the air. As long as the wave is traveling through a point, the air pressure at that point does not stay constant over time. The disturbance in an electromagnetic field is the change of electric and magnetic field. The wave can be looked upon as propagation of this disturbance.

As shown in Figure 1.1, you can describe a wave in terms of some parameters such as amplitude, frequency, and wavelength.

Figure 1.1 The Parameters of a Wave

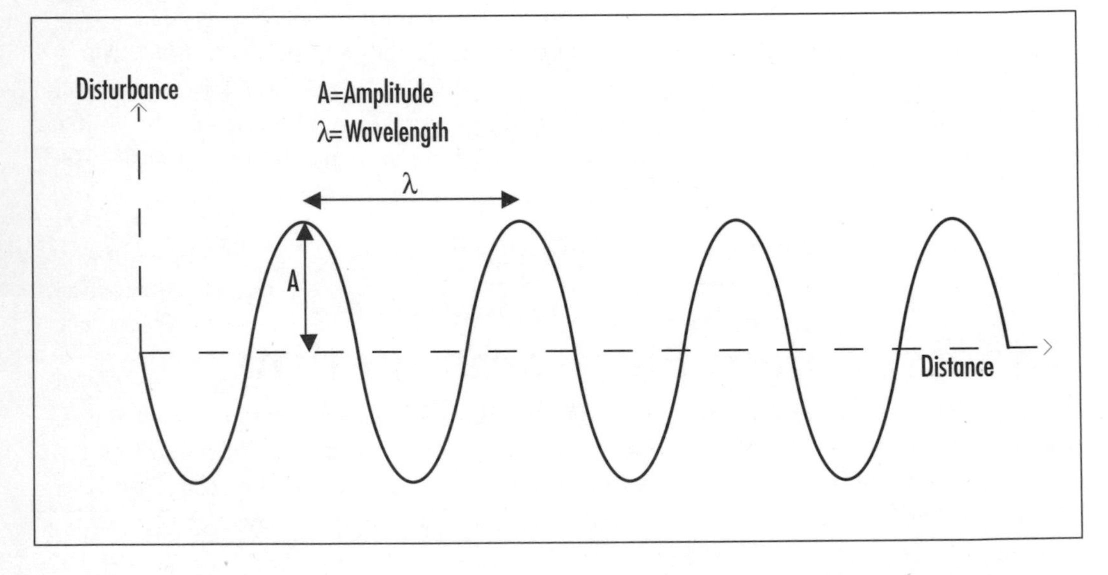

- **Wavelength** Denoted by the symbol λ, this is the distance between two consecutive crests or two consecutive troughs of a wave. The distance equal to wavelength makes one cycle of change.
- **Amplitude** Amplitude is the maximum amount of disturbance during one wave cycle.
- **Frequency** This is the number of cycles per unit of time a wave repeats. The frequency of an electromagnetic wave, *f*, propagating through free space (a vacuum), is calculated using the following equation:

  $f = c/\lambda$

  c is the velocity of light in vacuum. The frequency is measured in units of Hertz. One cycle per second is one Hertz, denoted by *Hz*.

- **Phase** This is the current position in the cycle of change in a wave.

So, what is the frequency of EM waves? EM waves cover a wide spectrum of frequencies, and the ranges of these frequencies constitute one way we define different types of EM waves.

Types of Electromagnetic Waves

Electromagnetic waves can be grouped according to the direction of disturbance in them and according to the range of their frequency. Recall that a wave transfers energy from one point to another point in space. That means there are two things going on: the disturbance that defines a wave, and the propagation of wave. In this context the waves are grouped into the following two categories:

- **Longitudinal waves** A wave is called a *longitudinal wave* when the disturbances in the wave are parallel to the direction of propagation of the wave. For example, sound waves are longitudinal waves because the change of pressure occurs parallel to the direction of wave propagation.
- **Transverse waves** A wave is called a *transverse wave* when the disturbances in the wave are perpendicular (at right angles) to the direction of propagation of the wave.

Electromagnetic waves are transverse waves. That means the electric and magnetic fields change (oscillate) in a plane that is perpendicular to the direction of propagation of the wave. Also note that electric and magnetic fields in an EM wave are also perpendicular to each other.

NOTE

Electric fields and magnetic fields (E and B) in an EM wave are perpendicular to each other and are also perpendicular to the direction of propagation of the wave.

Because electric and magnetic fields change in a plane (perpendicular to the direction of wave propagation), the direction of change still has some freedom. Different ways of using this freedom provide another criterion to classify electromagnetic waves into the following:

- **Linearly polarized waves** If the electric field (and hence the magnetic field) changes in such a way that its direction remains parallel to a line in space as the wave travels, the wave is called *linearly polarized*.

- **Circularly polarized waves** If the change in electric field occurs in a circle or in an ellipse, the wave is called *circularly* or *elliptically polarized*. Therefore, the polarization of a transverse wave determines the direction of disturbance (oscillation) in a plane perpendicular to the direction of wave propagation.

CAUTION

Only transverse waves can be polarized, because in a longitudinal wave, the disturbance is always parallel to the direction of wave propagation.

So, you can classify electromagnetic waves based on the direction of disturbance in them (polarization). The other criterion to classify EM waves is the frequency.

The Electromagnetic Spectrum

Have you ever seen electromagnetic waves with your naked eye? The answer, of course, is yes! Visible light is an example of electromagnetic waves. In addition to visible light, electromagnetic waves include radio waves, ultraviolet radiation, and X-rays (which of course are not visible to the naked eye). These different kinds of EM waves only differ in their frequency and therefore their wavelength. The whole frequency range of EM waves is called the *electromagnetic spectrum,* which is illustrated in Figure 1.2, along with the names associated with different frequency ranges within the spectrum.

Figure 1.2 The Electromagnetic Spectrum

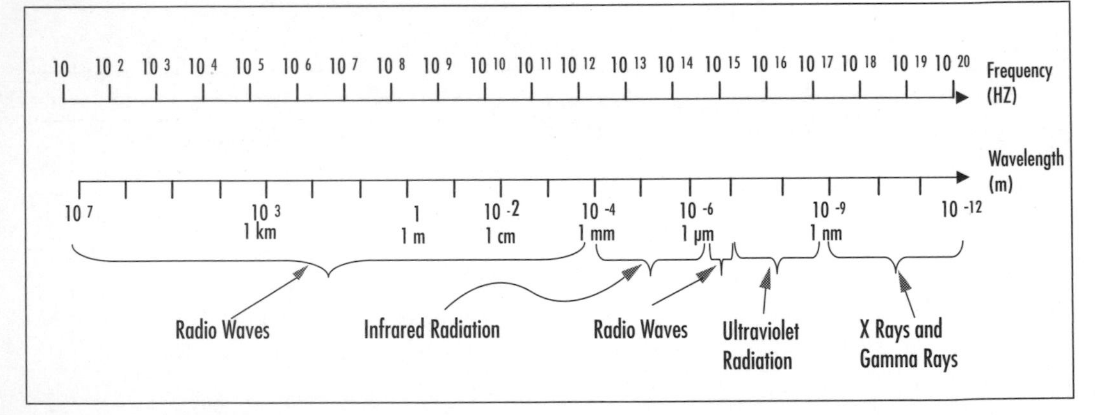

As shown in Figure 1.2, the radio waves occupy a major part of the electromagnetic spectrum. As the name suggests, a radio frequency identification (RFID) system uses radio waves to communicate.

If the numbers in Figure 1.2 do not make sense to you and if you have forgotten all about scientific notation, units of measurement, and logarithms, you will need to brush up on these math-related concepts to make your journey through this book smoother.

The Mathematics of RFID

This section discusses some math-related concepts such as scientific notation, units, and logarithm. Understanding these concepts will help you more firmly grasp the concepts discussed throughout this book.

Scientific Notation

To express numbers, scientists use a notation called *scientific notation*. It simplifies handling very large and very small numbers. Using this notation, you express a number as a product of a number between 1 and 10 and a power of 10. For example, the number 174,000 is expressed in scientific notation as:

1.74×10^5

To convert a number in scientific notation to the ordinary notation, here is the rule: Count as many places as the power of 10 after the decimal point, replace any empty place with a 0, and remove the point. For example:

$1.25 \times 10^4 = 12500$

$10^4 = 1 \times 10^4 = 10000$

Some powers of 10 have a name called a *prefix*. For example, 10^3 is called *kilo*, as in kilometer or kilogram. These powers of 10 in common use are shown in Table 1.1, along with the numbers they represent.

Table 1.1 Prefixes for Powers of 10

Power of 10	Number	Prefix	Abbreviation
10^{12}	1000,000,000,000	Tera	T
10^9	1000,000,000	Giga	G
10^6	1000,000	Mega	M
10^3	1000	Kilo	k
10^{-1}	1/10	Deci	d
10^{-2}	1/100	Centi	c
10^{-3}	1/1000	Milli	m
10^{-6}	1/1000,000	Micro	μ
10^{-9}	1/1000,000,000	Nano	n
10^{-12}	1/1000,000,000,000	Pico	p

NOTE

The power of 10 is also called *exponent*. For example, in 10^3, the number 3 is an exponent. In general, a mathematical operation written as x^n is called "*x* raised to the power *n*." This is also called *exponentiation*, with *x* as a base and *n* as an exponent.

In general, a^x is called an *exponential function*. It means *multiply the base with itself as many times as the exponent*. For example:

$2^3 = 2 * 2 * 2 = 8$

Remember the following two formulae for exponential functions. The first formula is:

$a^x * a^y = a^{x+y}$

For example:

$2^2 * 2^3 = 2^5 = 2*2*2*2*2 = 32$

The second formula is:

$a^x / a^y = a^{x-y}$

For example:

$2^5/2^3 = 2^2 = 2*2 = 4$

In addition to exponentiation, there is another function relevant to this book: the logarithmic function.

Logarithms

Logarithm is the inverse of an exponential function:

$y = a^x \Rightarrow x = \log_a y$

The expression $\log_a y$ is read as *log y to the base a*. For example:

$1000 = 10^3 \Rightarrow 3 = \log_{10} 1000$

The base 10 is a default for the term *log*; that is, *log (1000)* means *log of 1000 to the base 10*. After understanding the definition of log, you need to remember three more formulae for the log function. The first formula is:

$\log x^n = n * \log x$

For example:

$$\log 1000 = \log 10^3 = 3 * \log 10$$

The second formula is:

$$\log (x*y) = \log x + \log y$$

For example:

$$\log 1000 = \log(10*100) = \log 10 + \log 100$$

The third formula is:

$$\log (x/y) = \log x - \log y$$

For example:

$$\log 100 = \log (10000/100) = \log 10000 - \log 100$$

An example of use of your knowledge of logarithm is the decibel unit.

Decibel

Decibel, denoted by the symbol *db,* is a measure of the ratio of two values of a physical quantity such as power or voltage expressed in terms of logarithm. To be precise, the ratio X_1/X_2 of a physical quantity X will be expressed in decibels as:

$$X (db) = 10 * \log (X_1/X_2)$$

Exercise 1.2

How will the ratio of electric power be expressed in decibels in terms of the ratio of voltage?

Solution:

Recall that:

$$P = V^2/R$$

$$P (db) = 10 * \log(P_1/P_2) = 10 \log(V_1^2/V_2^2) = 10 \log (V_1/V_2)^2 = 2*10 \log (V_1/V_2)$$

$$= 20 * \log (V_1/V_2)$$

$$P(db) = 20 \log (V_1/V_2)$$

Now, if you see a relationship like this, you know why there is a 20 in front of *log* rather than 10.

Numbers in physics are used to express some quantities, and quantities are expressed in some kind of units.

Units

All physical quantities (except ratios) are measured in terms of basic amounts called *units*. The units for various physical quantities, along with the abbreviations commonly used, are presented in Table 1.2.

Table 1.2 Abbreviations for Units

Unit	Abbreviation	Unit of:
ampere	A	current
coulomb	C	charge
centimeter	cm	length
foot	ft	length
gram	g	weight
hour	h	time
hertz	Hz	frequency
inch	in	length
kilometer	km	length
meter	m	length
mile	mi	length
minute	min	time
millimeter	mm	length
millisecond	ms	time
nanometer	nm	length
ohm	Ω	resistance
pound	lb	weight
second	s	time
volt	V	voltage
watt	W	power
yard	yd	length

There are multiple systems of units. For example, length is expressed in miles in the customary U.S. system of units, whereas it is expressed in kilometers in the international system (IS) of units. Some conversions between these two systems relevant to the material in this book are presented in Table 1.3.

Table 1.3 Length in Two Different Units

U.S. Customary System Units	International System Units
1 in	2.54 cm
1 ft = 12 in	30.48 cm
1 yd = 3 ft	0.91 m
1 mi	1.61 km

Equipped with these basic physics and math concepts, you are now ready to explore the RFID field. Let's start by taking the bird's-eye view of the RFID landscape.

An Overview of RFID: How It Works

The story of RFID starts with one word: identification. RFID is here to replace existing identification technologies such as the barcode, which is used to identify an item by assigning it a unique number. An example of the barcode is shown in Figure 1.3. No doubt you have seen such barcodes on various products ranging from water bottles to wine cartons and from books to cases that contain quantities of items.

Figure 1.3 An Example of a Barcode on a Book

According to a display in the Smithsonian Institution's National Museum of American History, the first purchase of a product with a barcode was made on June 26, 1974, at a supermarket in Ohio. Today, almost everything that you buy from retailers has a barcode printed on it. These barcodes help manufacturers and retailers in the following ways:

- Keep track of inventory
- Provide information about the quantity of products being sold
- Speed up the checkout process

The barcode technology has the following limitations:

- A barcode identifies a type of product, not an individual item in that type.
- Tracking is not automatic. For example, to keep track of inventory, you must scan each barcode on every item of a product.
- A barcode does not contain much information other than the product type code.
- A barcode is a read-only technology; that is, you cannot change the information on the barcode or add new information to it.

So, the basic promise of barcodes is to provide identification of products at the class level. RFID is replacing those barcodes with a greater promise: automatic and global identification and tracking of objects (at the individual level), which could include almost anything: individual product items in retail stores, animals, trees—even people. Here is one of many possible scenarios relating how RFID works:

1. A label-like electronic device called a *tag* is attached to an object that needs to be identified and tracked. The tag contains the unique identification of the object and possibly more information about it.
2. Another electronic device called a *reader* is mounted at specific localities.
3. When a tagged object passes near any reader, the reader communicates with the tag and gets the information that the tag has about the object.
4. The reader passes the information to a host computer, which is typically part of a network connected to the Internet.
5. The host computers from several localities send the information about tagged objects to a central location.
6. The information is integrated at the central location into database management systems and can be analyzed by enterprise applications.

This scenario is depicted in Figure 1.4. The readers and tags use EM waves in the radio wave frequency range to communicate with each other.

NOTE

Note. A reader is also called an *interrogator,* and a tag is also called a *transponder*.

Figure 1.4 Readers Collect Information from Tags at Various Locations and Send It to a Central Location Over the Internet

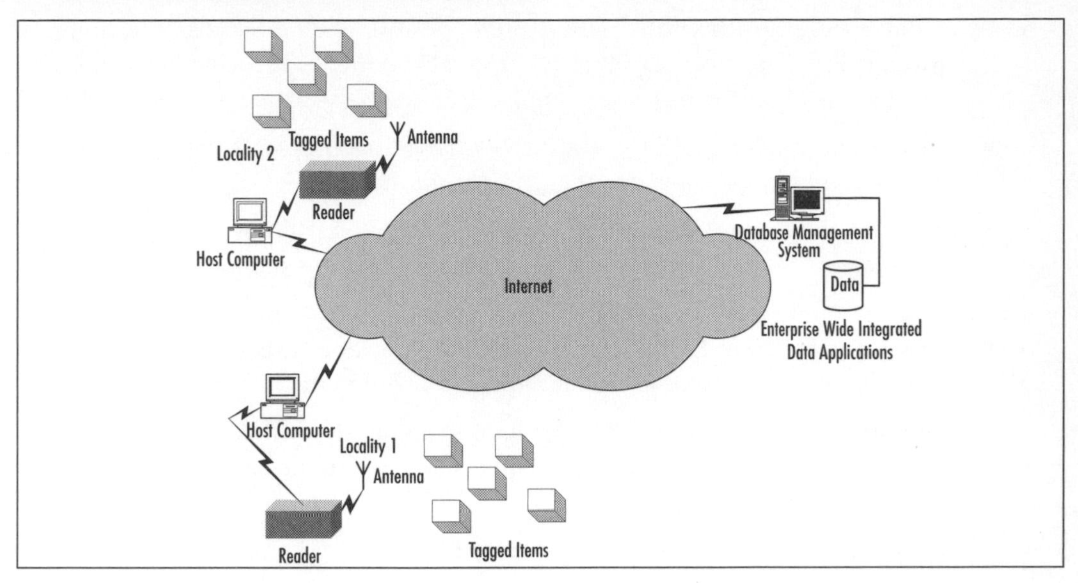

The advantages of RFID technology over barcode technology are as follows:

- The identification and tracking offered by RFID is at individual item level as opposed to the type level.
- A tag can contain more information about the object than just its ID.
- Depending on the type of tag, you can change the information on it.
- The objects can be tracked globally, automatically, and in real time, if needed.

In other words, an RFID tag attached to an object is an intelligent barcode that can communicate through readers to a global network system to inform it where the object is. RFID technology can support a wide spectrum of applications, from tracking cattle to tracking trillions of consumer products worldwide, thereby enabling manufacturers to know the location of each product during its life cycle, from the time it's manufactured to the time it's consumed and tossed in a recycle bin or a trash can. You can see that RFID is going to be more ubiquitous than barcode, and its applications are limited only by your imagination. Here is a list of some applications to get you started:

- **Asset tracking** This includes tracking of assets everywhere, such as in offices, labs, warehouses, and libraries.
- **Automated toll collection system** A reader on the highway toll booth and a tag attached to the vehicle's windshield facilitate automatic charging to the

car owner's account and eliminate the need for the driver to stop and manually pay the toll.

- **Health care applications** This includes positively identifying and tracking patients in a health care facility or a hospital, linking a patient with the right medicine and doctor or nurse, identifying unresponsive patients, and so on.
- **Livestock tracking** This includes tracking animals in places such as farms and zoos and linking them to their proper locations.
- **Supply chain tracking** This includes tracking items through the supply chain and managing inventory. The supply chain field is the key early adopter of RFID technology.
- **Tracking in manufacturing** This includes tracking parts during the manufacturing process as well as tracking the assembled items.
- **Tracking in retail stores** This includes tracking store trolleys and shelves, thereby facilitating automatic payment, checkouts, and inventory management.
- **Tracking in Warehouses** This includes real-time inventory tracking and management in a warehouse or storeroom by tracking items inside, items coming in, and items going out.
- **Tracking you** Yes, RFID will track any object, including people—for example, tracking people entering a certain area for security purposes, automatic contact management at events instead of sticking notes on bulletin boards, tracking babies in hospitals, tracking children at theme parks and festivals, and so on.

"Hold on—tracking *me*?" you say, and you'd be right about the privacy issues. But that's a topic for another book.

So the two main players in a core RFID system are the reader and the tag. You can start asking questions about them, such as this one: From how far apart can a reader and a tag communicate with each other? In other words, how large is the read range? Well, it could be anywhere from a centimeter to a few meters, depending on several factors, including the tag type and the value of the radio frequency being used for communication, called *operating frequency*.

Next, what do we mean by tag types? The tags can be categorized by different criteria. One of those criteria is the power source from which tags will draw energy to operate and to communicate. The tags that have their own power source such as a battery are called *active tags*, whereas the tags that do not have their own power source are called *passive tags.* A passive tag cannot do anything until it receives a signal (radio wave) from a reader to wake it up. It uses part of the energy from the signal to operate and the rest to communicate back to the reader—that is, to send back a radio wave. Recall the concept of inductive coupling, discussed earlier in this chapter. This is what goes on between a reader and an inductive passive tag: The magnetic energy is transferred from the reader to the passive tag through inductive coupling to power it up. It's as though the reader were saying, "Hello, Mr. Tag, time to wake up and tell me everything you know about this object."

Just like the read range, the readers and tags come in various sizes and shapes. Figure 1.5 shows a reader and a tag on the smaller end of the size spectrum. I know your next question: How do a reader and a tag really communicate with each other? That question goes to the physics behind RFID, which is discussed in the next chapter.

Figure 1.5 A Reader and a Tag: Skyetek's M1-mini *(Image courtesy of Skyetek)*

For now, note that neither the physics behind RFID nor the RFID technology itself is new. But it's only recently that greatness has been bestowed upon RFID by giant influencers such as the U.S. Department of Defense and Wal-Mart in their mandates and in a flurry of industrial mandates that followed. Now, armed with these mandates, government legislations, and the resulting hyperbole, RFID has set its journey to change the world. The forthcoming chapters will help prepare you to make your contribution to this revolution.

NOTE

Talking about legislation, the U.S. State Department has legislated that all U.S. passports must contain an RFID chip (tag) by the end of 2006. The chip, in addition to holding the standard passport data—name, address, birth date, and nationality—will also be able to hold biometric information such as iris scans and digital fingerprints. The European Union has its own RFID passport initiative under way.

The three most important takeaways from this chapter are the following:

- Electromagnetic force, one of the four basic forces that govern our universe, exhibits itself in the form of electromagnetic waves, which underline the physics behind RFID.
- While working with RFID, you will use simple mathematical concepts such as power of 10, logarithms, and some simple unit conversions.
- At the heart of an RFID system are two kinds of communication device: readers and tags. A tag is attached to an object that needs to be identified and tracked and contains information about the object. The reader collects the information about the object from the tag. Readers and tags use radio waves, a type of electromagnetic wave, to communicate with each other.

Summary

Our universe is governed by four natural forces: gravitation force, strong nuclear force, weak nuclear force, and electromagentic force. Where there is a force, there is energy, which is the ability of the force to do work. The amount of work done can be expressed in terms of power, which is the amount of energy transfer per unit of time. Work is performed when a force acts on an object and causes a change. For example, the Sun makes the Earth revolve around it by exerting gravitational force on it. Similarly, charged objects separated from each other can exert electromagnetic force on each other. How does an object exert force on another object without touching it? That happens through the field that exists between the two objects due to the force.

Of the four basic forces in the universe, the force that is relevant to RFID is the electromagnetic force, which exhibits itself in terms of electromagnetic waves. Electromagnetic waves, like any other wave, are characterized by their frequency and wavelength. These waves cover a wide spectrum of frequencies, called electromagnetic spectrum. Waves corresponding to one of the ranges in this spectrum are called radio waves. The radio waves are used by an RFID system for communication.

At the heart of an RFID system are two kinds of communication devices: tags and readers. A tag (an alternative to the barcode) is placed on an object that needs to be identified and tracked. The readers mounted at various locations read the information about the object from the tag and report it to the host computer, which in turn can send this information to a central location over the Internet. This way, an object can be tracked globally and in real time in an automatic fashion.

After learning the basic physics concepts in this chapter, you are ready to explore the physics behind RFID in the next chapter.

Chapter 2

RFID+

The Physics of RFID

Exam Objectives	What It Really Means
8.1 Identify RF propagation/ communication techniques.	Understand modulation, inductive coupling, and backscattering. You must know that low frequency systems use inductive coupling because low frequency means high wavelength means large antenna.
8.2 Describe antenna field performance/characteristics as it relates to reflective and absorptive materials (may use scenarios).	Understand the physical quantities that describe the loss of power or a change of direction in an RF wave propagating through the space, such as absorption, reflection, refraction, and scattering. You must know that water is a good absorbent of RF waves and metals are good reflectors.
8..3 Given a scenario, calculate radiated power output from antenna based on antenna gains, cable type, cable length, interrogator transmit power (include formulas in scenario).	Learn how to calculate physical quantities related to the power emitted by an antenna such as antenna gain and effective radiated power (ERP). You must also know what characteristics impact the propagation of the RF signal from the source to the antenna such as cable loss and impedance.

Introduction

The core functionality of an RFID system is the communication between a reader and a tag. The communication is carried out using RF waves, which are basically the EM waves with frequencies from the subspectrum of EM frequency spectrum called *radio frequencies*. The propagation of these waves is governed by the underlying physics principles. The goal of this chapter is to help you understand some physics concepts related to this communication. To accomplish this goal, we will explore three avenues: generation and propagation of the RF wave carrying the data signal from the source to the antenna, emission of the RF wave by the antenna into the free space, and propagation of the RF wave traveling through the space. Pay attention to the characteristics that affect the performance of an RFID system during this journey of the RF wave.

Understanding Radio Frequency Communication

Generally speaking, RFID is a means to identify an object using radio frequency transmission, which suggests that communication is involved in the identification process. The communication takes place between two devices: a reader that needs the information and a tag that has the information. Before we dive into the physics of communication, let's get on the same page about some concepts that are at the heart of this communication.

Elements of Radio Frequency Communication

Radio frequency communication uses the EM waves with frequencies from a specific part of the EM frequency spectrum. Therefore, the underlying physics behind RF communication is the same as for any communication that uses electromagnetic waves to carry information. The four major players that make this communication happen are the following:

- **Data signal** This is the wave that actually contains the information that needs to be sent to the receiver.
- **Carrier signal** This is the wave that carries the data signal.
- **Modulation** This is the process that encodes the data signal into the carrier signal and creates the radio wave that is actually transmitted by the antenna to propagate.
- **Antenna** This is a device used to transmit and receive signals such as radio waves.

NOTE

In an RFID system, both the reader and the tag have their own antennas through which they communicate with each other. A tag is also called a *transponder* because it responds to the reader's attempt to read it, and the reader is also called a *transceiver* because it receives information from the tag.

Here is how these four players work together to make the communication happen. First, understand that the information is communicated through changes (such as vibrations) in the carrier signal. The carrier signal itself is a constant signal unchanging in frequency and voltage—for example, a sine wave. It represents no information. As an analogy, I would not convey much information if I merely produced a constant sound out of my mouth, such as:

OOOOOOOOOOOOOOOOOOOOOOO

To convey some information, I would need to speak different sentences and different words in a sentence. In radio frequency communication, the information is encoded into the carrier signal using a technique called *modulation*, which means variation or change. You take the data signal that represent the information and impress it on a constant radio wave called a carrier. The data signal, as a result, varies (or modulates) the carrier wave. Once transmitted through an antenna, the two go together dancing over the air in the form of a modulated signal. The process of encoding the data signal into the carrier wave is called *modulation*. The transmitted modulated signal is received by the antenna on the receiving end and is demodulated to obtain the data signal. The process is depicted in Figure 2.1.

Figure 2.1 The Process of Communication Using Modulation

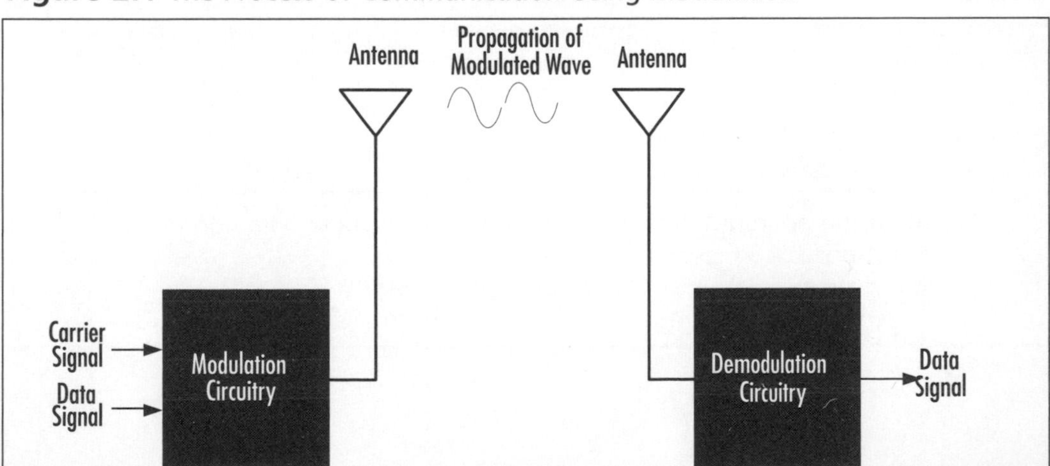

That all sounds good. But note that the original data signal itself has information in it, which is represented by the changes inherent in the signal. So the question is: Why don't we transmit the original data signal, or why do we need modulation in the first place?

Modulation: Don't Leave Antenna Without It

There are several reasons for the use of modulation in communication. Discussing the following two will be sufficient for the scope of this book.

The Propagation Problem

A data signal generally comprises a whole range of different frequencies together. The problem with the low-frequency components of the signal is that few communication media will allow the propagation of low frequencies without distortion. Modulation presents the solution to this problem by copying these low-frequency components to high-frequency carrier waves.

The Transmission Problem

The low-frequency data signal will have a high wavelength and as a result will require very large antennas for transmission and reception. Here is the rule of thumb: To achieve a useful amount of radiation, the antenna length should be at least one quarter of the wavelength of the wave to be propagated. For example, consider a signal component with frequency of 1 KHz. The wavelength for this wave will be:

$\lambda = c/f = (3 \times 10^8 \text{ m/s})/ (10^3 \text{ 1/s}) = 300 \text{ km}$

Antenna length = $\lambda/4$ km

A 75-kilometer-high antenna (the tower of Babylon)? You get the point. Modulation solves this problem by sending the low-frequency signal inside a high-frequency carrier wave. For example, as shown in Exercise 2.1, a frequency of 900 MHz (for the carrier signal) will give you the antenna length of 8.3 cm.

Exercise 2.1

Calculate the optimum antenna size for a cellular phone working at 900 MHz.

Solution:

$\lambda = c/f = (3 \times 10^8 \text{ m/s})/ (900 \times 10^6 \text{ 1/s}) = 1/3 \text{ m}$

Antenna length = $\lambda/4$ m = 1/12m = 8.33 cm

So, basically, the modulation promotes the frequency band (range of frequencies) of the data signal to higher-frequency bands so that it can be transmitted for a safe journey. It's important to understand the role of these frequency bands in the story of radio frequency communication.

Frequency Bands in Modulation

In the description of the modulation used for communication, some terms referring to different frequency bands are often used. A *frequency band* refers to a specific range of frequencies. These terms are described as follows:

- **Baseband** This is the range of frequencies of the original data signal before modulation.
- **Sideband** This is the frequency band on either the higher side or the lower side of the carrier frequency band within which the frequencies produced by modulation fall.
- **Upper sideband (USB)** This is the sideband above the carrier frequency.
- **Lower sideband (LSB)** This is the sideband below the carrier frequency.

As depicted in Figure 2.2, the information (data) is carried in the sidebands. In Chapter 1, you learned about the components of a wave: amplitude, frequency, and phase. You can ask a question now: For which of these components does the modulation vary (or change)? The answer is that you can change any of these components and accordingly there are several modulation techniques or types.

Figure 2.2 The Sidebands in Modulation That Carry the Information

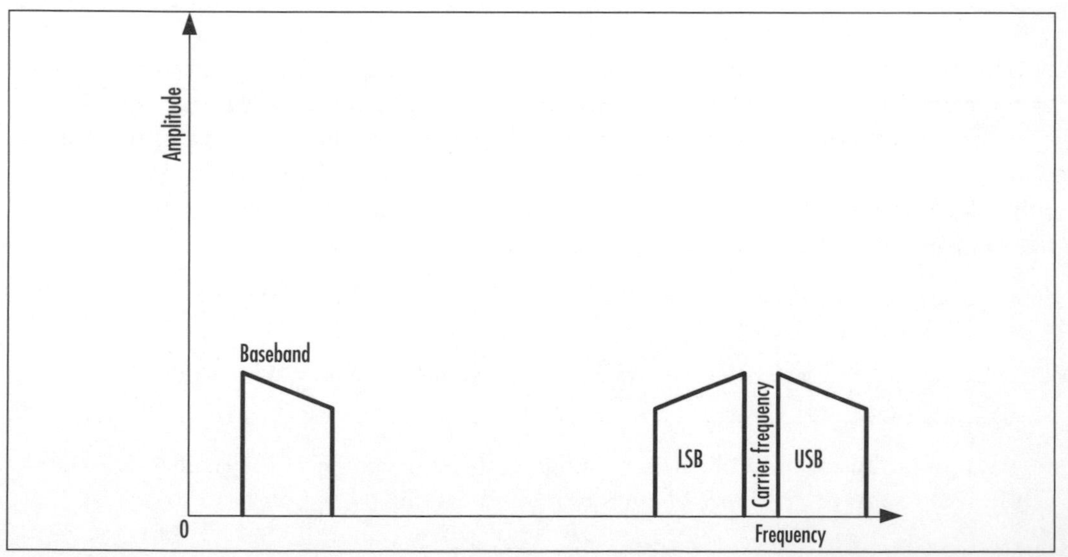

Understanding Modulation Types

Depending on which component of the wave is changed to encode data, there are different types of modulation, such as amplitude modulation, frequency modulation, and phase modulation.

Amplitude Modulation and Amplitude Shift Keying

Amplitude modulation (AM) is the technique in which the amplitude (peak-to-peak voltage) of the carrier wave is varied as a function of time in proportion to the strength of the data signal. As shown in Figure 2.3, the originally constant amplitude of the carrier signal rises and falls with each high and low of the data signal.

Figure 2.3 Loading the Data Signal on a Carrier

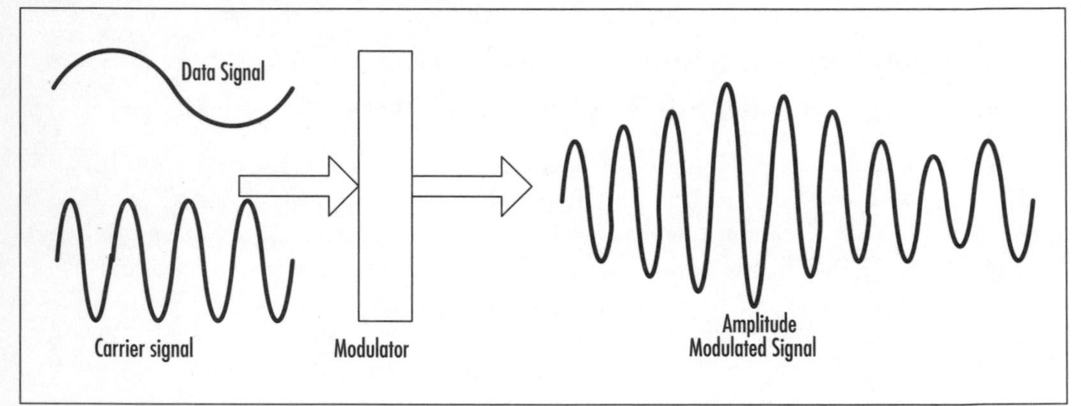

In its basic form, amplitude modulation produces a signal with power concentrated at the carrier frequency and in two adjacent sidebands. Each of the two sidebands is equal in bandwidth to that of the modulating signal and is a mirror image of the other. This type of AM is called *double sideband full carrier (DSBFC),* meaning that you use the full power of both the sidebands and the carrier for transmission. This type is also called *double sideband (DSB)*. There are two problems with this picture:

1. Only one of the two identical sidebands is needed; the other one is just a waste of power.
2. Half the power is concentrated at the carrier frequency, which carries no useful information.

The solution to this problem is to suppress the carrier, one of the sidebands, or both. This gives rise to several types of amplitude modulation:

- **Double-sideband reduced carrier transmission (DSB-RC)** This type of modulation uses full power of both sidebands but reduces the carrier level (amplitude). To be precise, this is the type of AM achieved by implementing the following two requirements:
  - The frequencies produced by the modulation are symmetrically spaced above and below the carrier frequency.
  - The carrier level is reduced for transmission at a fixed level below, which is below the level of the carrier provided to the modulator.
- **Double-sideband suppressed-carrier transmission (DSB-SC)** This is a special case of DSB-RC and is achieved by implementing the following two requirements:
  - Frequencies produced by amplitude modulation are symmetrically spaced above and below the carrier frequency.
  - The carrier level is reduced to the lowest practical level; ideally speaking, it's completely suppressed.
- **Single-sideband (SSB)** This is the type of AM in which only one sideband and the carrier is used. You can also call it *single-sideband full carrier.* Note that it is not necessary to transmit both sidebands: Either one can be suppressed at the transmitter without any loss of information. The advantages of SSB include smaller transmitter power, smaller bandwidth (one-half that of a DSB), and less noise at the receiver.
- **Single-sideband suppressed-carrier (SSB-SC)** This is the SSB in which the carrier is suppressed. Even greater efficiency is achieved this way by completely suppressing both the carrier and sideband. This modulation type is widely used in amateur radio due to its efficient use of both power and bandwidth.

The less power to be transmitted by these AM types results in significantly less size, weight, and peak antenna voltage requirements of the SSB transmitter than those for the standard AM transmitter.

NOTE

In DSB-RC modulation, the carrier level is selected so that it is suitable for use as a reference by the receiver, except for the case in which it is reduced to the minimum practical level—for example, the carrier is suppressed.

The amplitude modulation is called *amplitude shift keying (ASK),* when the data signal is a digital signal. The term *keying* is the legacy term from the times of telegraphy, when an operator would manually push keys to make short and long tones. The kind of keying we're interested in refers to which characteristic of the analog carrier signal is to be varied to represent the ones and zeros of a digital data signal: amplitude, frequency, or phase.

ASK varies the amplitude of a carrier signal to represent binary data. The binary information is transmitted by assigning discrete amplitudes to bit patterns. For example, Figure 2.4 presents a simple example of ASK by showing the modulated signal corresponding to the digital signal that represents the binary number 0011010. Note that in the modulated signal, the period is the same for the entire signal; only the amplitude varies. In this example, an amplitude of 1 represents a 0, and the amplitude of 2 represents 1.

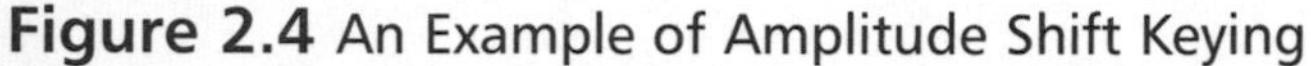
Figure 2.4 An Example of Amplitude Shift Keying

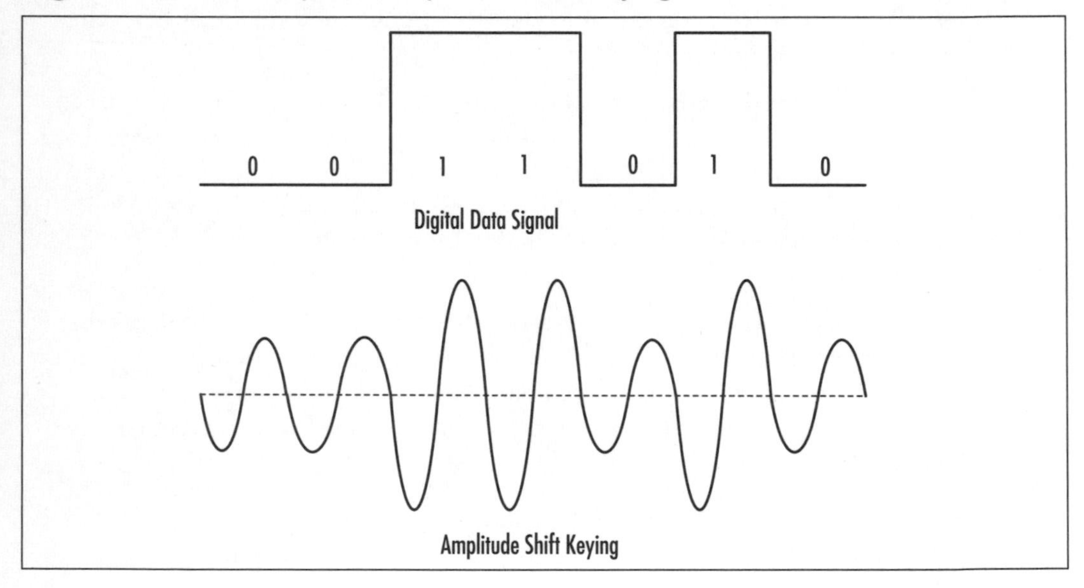

You can also encode data into the carrier signal by varying frequency instead of amplitude.

Frequency Modulation and Frequency Shift Keying

Frequency modulation (FM) is the modulation technique that represents information as variations in the frequency of the carrier wave, whereas in AM, the carrier amplitude is varied while its frequency remains constant. In analog applications, the carrier frequency is varied in direct proportion to changes in the amplitude of the data signal, as shown in Figure 2.5.

Figure 2.5 An Example of Frequency Modulation

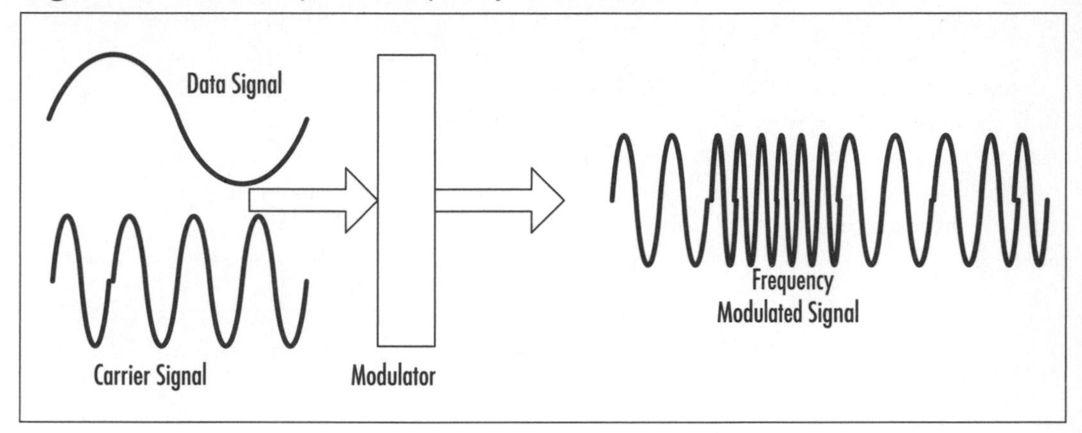

If the data signal is a digital signal, the FM technique is called *frequency shift keying*. In this case, the digital data is represented by shifting the carrier frequency among a set of discrete values.

NOTE

FM is most commonly used to transmit signals at VHF, whereas AM is most commonly used for transmitting audio signals (LF).

So, FSK modulates the frequency of the carrier signal to represent data. The binary information is transmitted using different frequencies to represent bit patterns: one frequency represents one binary bit and a different frequency represents the other binary bit. Obviously, these frequencies lie within the bandwidth of the transmission channel.

Figure 2.6 presents a simple example of a modulated signal using FSK. The signal is representing the binary number 0011010.

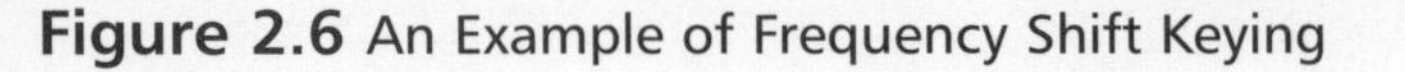

Figure 2.6 An Example of Frequency Shift Keying

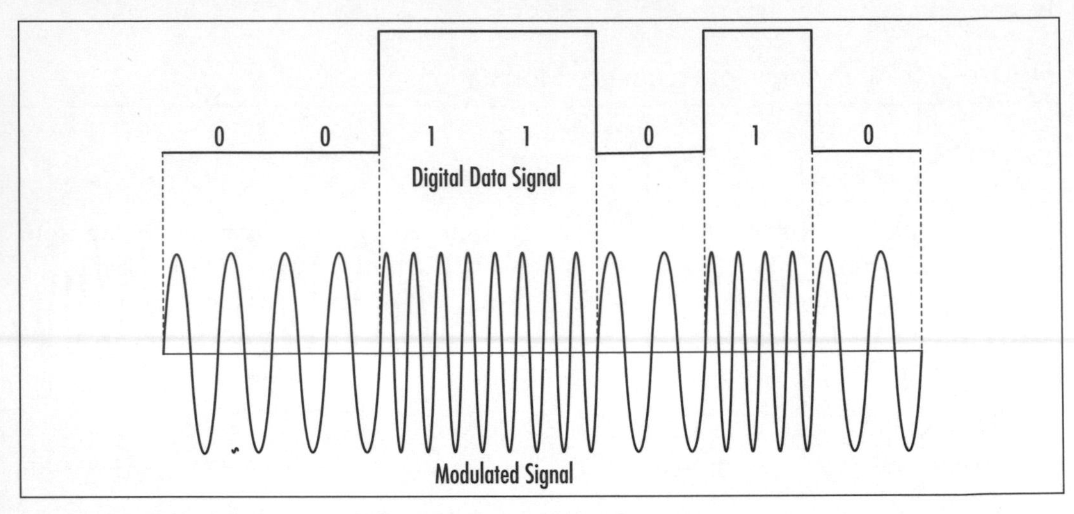

If, instead of varying amplitude or frequency, you vary the phase of the carrier wave to encode data signal, you are using phase modulation.

Phase Modulation and Phase Shift Keying

Phase modulation (PM) is that kind of modulation in which information is represented by variations in the phase of the carrier wave. Unlike AM and FM, PM is not very widely used. When the data signal is a digital signal, the corresponding phase modulation technique is called *phase shift keying*.

So, phase shift keying is a technique that represents digital data by shifting the period of the carrier signal. The binary information is transmitted by assigning different phases to different bit patterns. Figure 2.7 presents a simple example of phase shift keying.

Figure 2.7 An Example of Phase Shift Keying

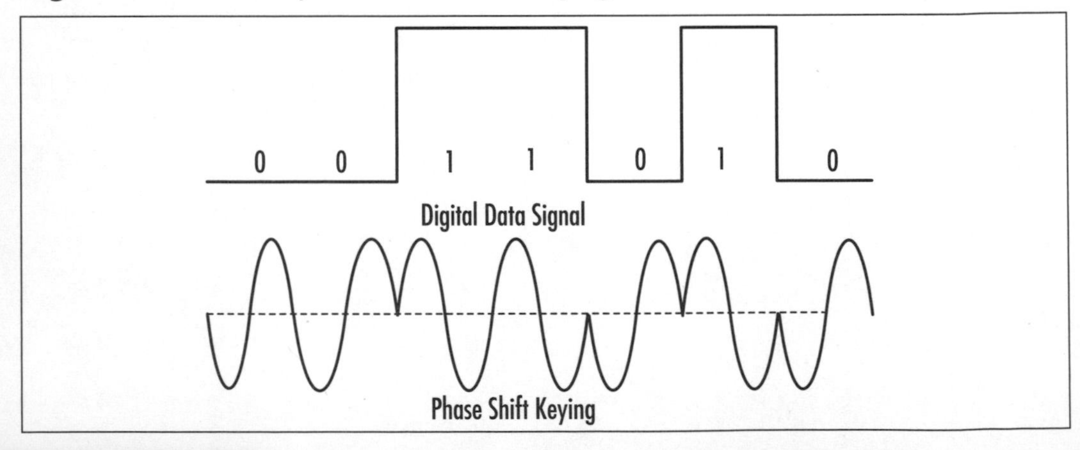

The binary signals can be represented in ways simpler than amplitude shift keying, frequency shift keying, or phase shift keying. After all, we are only talking about ways to represent two states: 1 and 0, or on and off.

On-Off Keying (OOK)

On-off keying (OOK) is a type of modulation in which the digital data is represented as the presence or absence of a carrier wave. For example, the presence of a carrier for a specific duration represents a binary one, whereas the absence of the carrier for the same duration represents a binary zero. This technique is most commonly used to transmit Morse code over radio frequencies. It has also been used in the industrial, scientific, and medical (ISM) radio bands to transfer data between computers. Figure 2.8 presents a simple example of on-off keying.

Figure 2.8 An Example of On-Off Keying

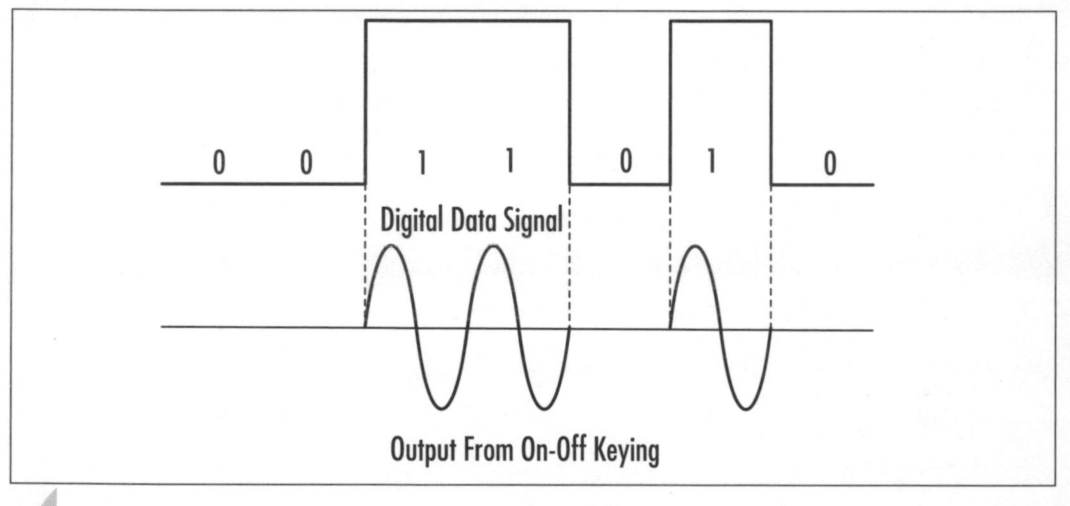

NOTE

ISM bands are the bands of radio frequencies originally reserved internationally for noncommercial use for industrial, scientific, and medical purposes.

In this section, we explored the techniques to encode data into the carrier signal. Now the question is: How is the signal carrying the information transferred from the sender to the receiver?

RFID Communication Techniques

Communication is basically the transfer of information—that is, to send information from one location and to receive it at another. In the RF world, this is accomplished by the transfer of energy (which contains the information coded in it) through RF waves. There are two main communication techniques that the RFID readers and tags use to communicate with each other. These techniques are *coupling* and *backscattering*.

Communication Through Coupling

Coupling, in general, is the transfer of energy from one medium, such as a metallic wire or an optical fiber, to another similar medium. Some examples of coupling include capacitive (electrostatic) coupling and inductive (magnetic) coupling.

As explained in Chapter 1, inductive coupling is the process of transferring energy from one circuit to another through a shared magnetic field by virtue of the mutual inductance between the two circuits. Note the following points about inductive coupling:

- Inductive coupling is used by low-frequency or high-frequency RFID systems. This way the tag and the reader can use a loop-style coil for an antenna because the traditional antenna would need to be too long due to the long wavelengths of the low-frequency waves.
- Inductive coupling works only in the near field of the RF signal.
- Sometimes inductive coupling is further subdivided into two kinds of coupling:
    - Close coupling within a range of about 1cm
    - Remote coupling within a range of about 1 cm to 1m

The power transfer between the two coils depends on the following quantities:

- Operating frequency of the system
- Number of turns/windings in the coils
- Area enclosed by each coil
- Angle the coils make with each other; for maximum power transfer, the coils should be aligned in the same plane
- Distance between the two coils

The magnetic field can be used to transfer energy only within the short range. For long-range communication, you need to send information through EM waves (radiation). This technique used in RFID systems is called *radiative coupling* or *backscattering*.

Communication Through Backscattering

Backscattering is the process of collecting an inbound signal (energy), changing the signal (the data it carries), and reflecting it back to where it came from. The long-range RFID systems operating at ultra-high frequency (UHF) or microwave frequencies use this communication technique. The reader sends out the information in the form of an EM wave at a specific frequency; the tag receives the wave, encodes the information into the wave (changes the wave), and scatters it back to the reader.

When you design and install a system, there is always a set of performance requirements that could differ from customer to customer. The antenna is an important component of an RFID system. Therefore it's important to understand what constitutes and affects the performance of an antenna.

Understanding Performance Characteristics of an RFID System

Radio devices communicate using antennas for transmitting and receiving the signals. Just like any other radio device, RFID tags and readers can also communicate with each other using antennas. The information is encoded into an RF wave and sent to the antenna through a transmission line. So, antennas play a vital role in an RFID system, and it is important to understand the characteristics of the transmission line and antennas that impact the performance. These characteristics are discussed in the following sections.

Cable Loss

RFID systems typically use 50-Ω coaxial cable as a transmission line. Cable loss is the amount of signal power lost in the cable. The longer the cable, the greater the loss.

Impedance

Impedance is defined as resistance to the flow of current in a circuit element and is measured as a ratio of voltage, say V, across the element and current, say I, through the element:

$$Z = V/I$$

The antenna receives power (in terms of current) from the source through the transmission line. The input impedance, Z_i, for the antenna is the following:

$$Z_i = V_i/I_i$$

V_i is the antenna input voltage, and I_i is the antenna input current.

To realize how impedance affects performance, note that the electromagnetic wave (power) travels through different parts of the antenna system, which can have different values for impedance. The parts to be considered here are the source that produces the power, the transmission line that brings the power to the antenna, and the antenna transmitter that transmits the power. The following are the different kinds of impedance defined in this case:

- **Characteristic impedance** This is the impedance of the transmission line, which is assumed to be lossless and of infinite length:

  $$Z_0 = (\mu/\varepsilon)^{1/2}$$

  where μ is the magnetic permeability of the medium that makes the transmission line and ε is the electric permeability of the medium. An example of transmission line is the antenna cable.
- **Antenna input impedance** The ratio of the input antenna voltage to the input antenna current.
- **Transmitter output impedance** The impedance used by the antenna's transmitter to transmit the power into the free space.

To get the best performance, it is important that all these impedances belonging to the different parts of an RFID system match with each other. If the antenna input impedance and the transmitter output impedance match the characteristic impedance of the transmission line, the antenna will radiate maximum power. However, there is always some impedance mismatch—for example, due to discontinuities in the transmission line or if the transmission line is terminated with other than its characteristic impedance. The impedance mismatch results in reflecting part of the wave energy back to the source and thereby impeding the performance. This phenomenon can be understood in terms of the voltage standing wave ratio.

The Voltage Standing Wave Ratio

A *standing wave*, also called a *stationary wave*, is a result of interference between two waves moving in the opposite direction. In an RFID system, this situation can arise due to the impedance mismatch along the transmission line from source to antenna transmitter. The impedance mismatch will result in reflecting part of the energy from the antenna back to the source, and the forward wave and the reflected wave will interfere with each other. Two cases for this interference are constructive and destructive, respectively:

- **Constructive interference** This is the case when the crests of one wave coincide with the crests of the other wave, and therefore the amplitude of the resultant wave is the sum of the amplitudes of the interfering waves:

$V_{max} = V_f + V_r$

- **Destructive interference** This is the case when the crests of one wave line up with the troughs of the other wave, and therefore the amplitude of the resultant wave will be the difference of the amplitudes of the interfering waves:

$V_{min} = V_f - V_r$

The *voltage standing wave ratio,* or VSWR, is measured as:

$VSWR = V_{max} / V_{min} = (V_f + V_r)/(V_f - V_r) = (1+ V_r/ V_f)/(1+ V_r/ V_f) = (1+\rho)/(1-\rho)$

where $\rho = V_r/ V_f$ is the magnitude of what is called the *reflection coefficient.*

A perfect impedance match will result in a VSWR of 1:1, which is practically impossible. VSWR is always expressed with 1 as the denominator.

EXERCISE 2.2

What is the value of VSWR for a short circuit and for an open circuit?

Solution: In both cases, the VSWR = infinity/1.

So, VSWR is the ratio of maximum voltage to minimum voltage along the transmission line in a standing wave situation. Another characteristic that can affect performance is noise.

CAUTION

The impedance and VSWR are considered during the manufacturing process to produce the desired output according to the standards and regulations. Any adjustments made to the cable or antenna can cause the change in VSWR and in the transmitted power and may violate the standards.

Noise

Noise is an unwanted electrical wave (or energy) present in a circuit or a signal. It is called noise to the signal or a background. The effect of the noise is represented by a quantity called *signal-to-noise ratio (SNR)* and can be calculated as shown here:

$$SNR = (A_s/A_n)^2$$

In this formula, A_s is the amplitude of the signal wave and A_n is the amplitude of the noise wave. SNR is usually represented in decibels:

$$SNR\ (dB) = 10 \log (A_s/A_n)^2 = 20 \log (A_s/A_n)$$

This equation tells us that when the noise is stronger than the signal, the value of SNR will be negative, in which case reliable communication is not possible unless we either increase the signal strength or decrease the noise.

Regardless of how strong the signal is compared to the noise, it's useless unless the receiver receives it. Polarization is a characteristic that you should know in this context.

Beamwidth

As shown in Figure 2.9, the *beamwidth* of an antenna is the angle between the two half-power points around the point (the main lobe) that has the peak effective radiated power.

Figure 2.9 An Example of Beamwidth

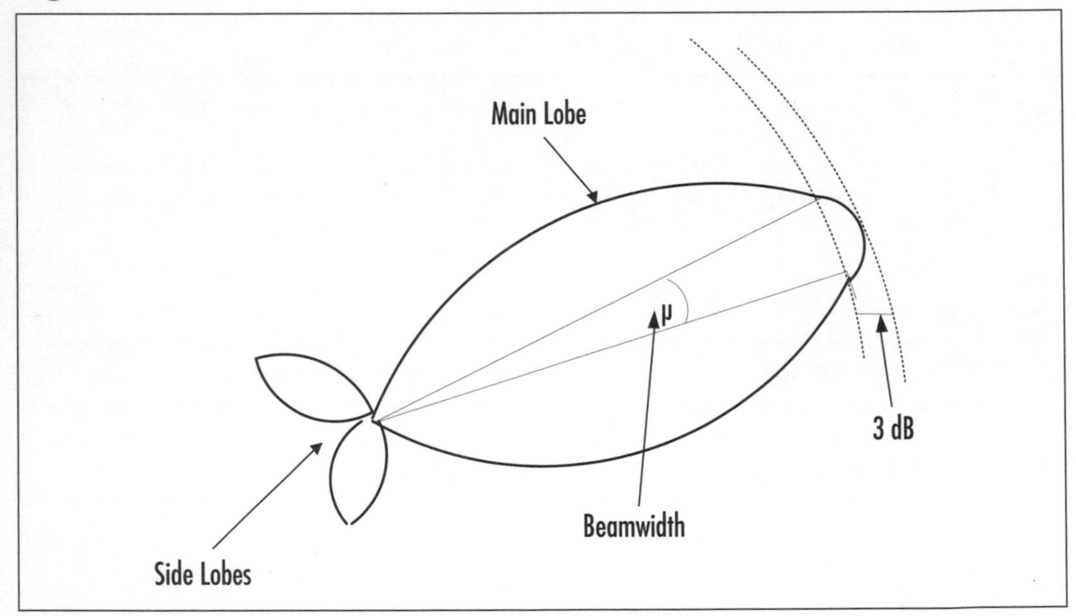

Exercise 2.3

Show that each of the two half-power points of the beamwidth are 3dB below the point of maximum radiation.

Solution: By definition, a half point has half the maximum radiation power. Therefore, the relative signal strength at a half-point can be expressed in decibels, as in the following:

$10 \times \log(1/2) = -3$ dB

Why is beamwidth important? RFID systems are not broadcast systems. A reader wants to get information from a tag at a specific location. So, the beam nature (focusing) of the radiated energy is important from the perspective of performance. We can talk about this issue in terms of directivity as well.

Directivity

The *directivity* of an antenna is defined s its ability to focus in a particular direction to transmit or receive energy. Directivity is calculated as the ratio of the maximum value of power transmitted (or received) per unit of solid angle to the average power transmitted (or received) per unit of solid angle. This property is important in an RFID system because the communication in this case is point to point between a tag and a reader, as opposed to broadcast, as in the case of FM radio. Directivity is a performance characteristic in an RFID system because the performance depends on how well the reader and tag can direct their energy at each other.

Note

In an environment of poor directivity, a reader may end up reading a tag outside its zone. This is called a *phantom read* or a *ghost read*.

Antenna Gain

Antenna gain is another way of measuring an antenna's ability to radiate in a specific direction. This is measured as a ratio of energy radiated at a point of maximum radiation to energy radiated at the same point by some reference antenna:

$A_g = P_{out}/P_{ref}$

One of the theoretical antennas used as a reference is called an *isotropic* (omnidirectional) antenna—that is, it radiates power uniformly in all directions. Therefore, the power radiated by the reference antenna can be taken as equal to the input power, assuming that the antenna is lossless. This would result in the following equation for the antenna gain:

$$A_g = P_{out}/P_{in}$$

Antenna gain is usually expressed in decibels:

$$A_{g\,(dB)} = 10 \times \log(P_{out}/P_{in})$$

EXERCISE 2.4

What is the antenna gain when the reference antenna is a dipole antenna?

Solution: With a dipole antenna as a reference, the maximum output power will be twice the input power. Therefore, the antenna gain can be calculated as:

$$A_g\ (dB) = 10x\log(2/1) = 3\ dB$$

So, antenna gain is important because in an RFID system the power is transmitted in preferred directions and is not broadcast uniformly in all directions. For example, a reader wants to direct the power at the tag it wants to read. That's why directivity and antenna gain are performance-related characteristics of antennas in RFID systems.

CAUTION

Don't be misled by the term *gain*. The overall output (transmitted) power is not greater than the overall input power. Only the output power in a specific direction is greater than some reference power in that direction. In other words, the antenna does not act as an amplifier.

Before an antenna can transmit power, it receives that power from the source through a transmission line. The characteristics of that transmission line (or the circuitry) are also important from the perspective of performance. One of those characteristics is impedance.

Polarization

As described in Chapter 1, the *polarization* of a transverse wave, such as an electromagnetic wave, refers to the direction of oscillations in the plane perpendicular to the direction in which the wave travels. The antenna of an RFID system emits electromagnetic waves into the free space. The polarization of the antenna refers to the direction of oscillations in these waves.

Based on polarization, there are two types of antenna:

- **Linearly polarized antennas** Linear polarization is relative to the surface of the earth. It is of two kinds: horizontally polarized waves travel parallel to the surface of the earth, whereas vertically polarized waves travel perpendicular to the surface of the earth.
- **Circularly polarized antennas** A circularly polarized wave basically spins as it travels.

Polarization is a performance characteristic because the readability of the tag greatly depends on the polarization of the antenna and the angle the tag makes with the reader. Here is how polarization affects performance:

- For a maximum transfer of power, the reader and the tag antennas should have the same polarization.
- If the transmitting antenna is horizontally polarized and the receiving antenna is vertically polarized (or vice versa), not much power transfer is going to happen.
- If the receiving antenna is circularly polarized, it will receive some radiation, regardless of the polarization of the transmitting antenna. This is because a circular polarization has both components of the linear polarization: horizontal and vertical.

The transmitter emits the energy (which contains the information) at a certain frequency, and the receiver that receives this energy is also tuned to a certain frequency. The performance can be optimized if the transmitter and the receiver resonate with each other.

Resonance Frequency

Due to the underlying physics principles, a system absorbs maximum energy when the frequency of the energy waves matches the system's own natural frequency, the *resonant frequency*. Matching means the system's frequency is the same as or an integral multiple of the frequency of the energy that's being received. For optimal performance, it is important that the receiving antenna in an RFID system match the frequency of the incoming field—that is, it needs to resonate with the frequency of the incoming field.

Typically, an antenna is tuned for a specific frequency that matches the frequency of the incoming field, called *resonant frequency* or the *base frequency*. The integral multiples of this base frequency will also be effective frequencies for the antenna.

For example, if an antenna is tuned for a resonant frequency f_r, it will be effective for frequencies such as $1f_r$, $2\ f_r$, $3\ f_r$, and $4\ f_r$. Because frequency is inversely proportional to wavelength, the corresponding effective wavelengths will be λ, $\lambda/2$, $\lambda/3$, and $\lambda/4$. These wavelengths are also called the *electrical length* of the antenna and make their way into the antenna design as the antenna size.

NOTE

The low- and high-frequency tag antennas will need to be very large to resonate with the operating frequency. This is why these tags are designed to work on the principle of inductive coupling.

All the quantities discussed in this section directly or indirectly refer to the amplitude, voltage, or energy, all of which affect the power an antenna will radiate or absorb. This is because communication between two radio devices such as a reader and a tag is carried out by exchanging power between the antennas of the two devices. Therefore, it's important to understand the physical quantities related to the power emitted by an antenna.

Performing Antenna Power Calculations

To understand an antenna's performance, it's important to know how an antenna radiates power. For example, an isotropic (omnidirectional) antenna radiates power uniformly in all directions, whereas a directional antenna radiates power in a specific direction. The performance of an antenna related to the power it radiates can be understood in terms of three physical quantities: effective radiated power, power density, and link margin.

Effective Radiated Power

The *effective radiated power (ERP)* of an antenna in a specific direction is the power that will need to be supplied to a reference antenna to produce the same power this antenna is producing in this direction. Therefore, by definition of antenna gain, the ERP can be written as:

$$ERP = P_t \times A_g$$

where A_g is the antenna gain and P_t is the total power transmitted by the antenna, which can be expressed in the following equation:

P_t = RF power – cable loss

EXERCISE 2.5

The ERP of an antenna system is 100,000 watts, and the antenna gain is 7 dB. How much total power does the antenna actually transmit?

Solution:

$A_g = 10 \times \log(ERP/P_t) = 7$ dB

Therefore:

$ERP/P_t = 10^{0.7} = 5 \Rightarrow P_t = ERP/5 = 20,000$ Watts

NOTE

Sometimes in the literature, you will come across the term *equivalent isotropic radiated power (EIRP)*. This term refers to the effective radiated power calculated by using the antenna gain, which was computed using the isotropic antenna as the reference antenna. In this case, ERP will be the effective radiated power calculated by using the antenna gain. which was computed by using the half-wave dipole antenna as the reference antenna. The relationship between EIRP and ERP can be shown by the following equation:

EIRP (W) = 1.64 x ERP (W)

This equation assumes a lossless transmission line between the transmitter and the antenna.

After power is transmitted by an antenna, it spreads out into the space. Therefore, the power density (power per unit space) is an important quantity.

Power Density

An EM wave transmitted from an antenna travels in all directions in the form of an expanding spherical *wavefront*. The *power density* can be looked upon as the power of this wave per unit of surface area of the sphere. The surface area of a sphere with radius R is $4\pi R^2$. Therefore, the power density, P_d, at a distance R from the transmitter antenna can be calculated using the following formula:

$$P_d = P_t/(4\pi R^2)$$

P_t is the total power radiated by the antenna. This formula works for the power being emitted by an isotropic antenna. If the antenna is a directional antenna, we need to take into account the antenna gain, and the formula used to calculate the power density is as follows:

$$P_d = ERP/(4\pi R^2) = (P_t \times A_g)/(4\pi R^2)$$

Once the antenna has radiated energy, bad things, in addition to the natural spreading out, can happen to it while it's on its way to the destination. For example, it may be absorbed or reflected back by some materials on its way. ERP does not account for what happens to the energy wave on its way to the destination and how it is received by the receiving antenna. However, the overall system performance depends on how much power is being transferred between the transmitter and the receiver. The quantity that includes the travel and the receiving part of communication is called *link margin*.

Link Margin

Link margin quantifies the performance of the overall RFID communication system, including the transmitting antenna and the receiving antenna. The link margin, L_m, can be defined as:

$$L_m = (ERP_r/P_{min}) = (ERP_t \times A_{rg})/P_{min} = (P_t \times A_{tg} \times A_{rg})/P_{min}$$

$$Lm\ (dB) = 10 \times \log((P_t \times A_{tg} \times A_{rg})/P_{min}) = 10 \times (\log P_t + \log A_{tg} + \log A_{rg} - \log P_{min})$$

where:

P_t = Transmitted power

A_{tg} = Gain for transmitter antenna

A_{rg} = Gain for receiving antenna

P_{min} = Minimum received signal strength

Looking at this equation, you can realize that link margin is the ratio of the maximum effective signal strength received to the minimum signal strength received. In RFID, it means the amount of power that a tag can extract from the RF signal before the communication between the tag and the reader weakens.

So, the link margin takes into account the impacts of both the transmitting antenna and the receiving antenna. It also includes the factor of minimum received signal strength. The received signal strength varies and is less than the transmitted signal strength due the interaction of the signal with the medium through which it travels.

The Travel Adventures of RF Waves

When an RF wave travels from the transmitter to the receiver, it can be affected by various factors discussed in the following sections.

Absorption

When an RF wave strikes a material object, some of its energy will be absorbed by the object, depending on the frequency of the wave and the material of the object. Water and objects containing water, such as liquid products, wood, and food, are especially good at absorbing RF waves. UHF waves, due to their shorter wavelengths, are more susceptible to absorption than LF and HF waves.

Attenuation

Attenuation in general means a decrease in the amount of something. In RF physics, it means the decrease in amplitude (strength) of the RF signal (wave). Attenuation is the opposite of amplification. It can occur when the signal is traveling from the source to the antenna through the transmission line or during propagation from the transmitter antenna to the receiver antenna. It can occur due to a number of reasons, such as absorption and dispersion.

Dielectric Effects

Dielectric effects refer to a medium's capacity to retain charge. As a result, an electromagnetic wave traveling through a dielectric medium is slowed down. The strength of this effect is measured by a quantity called the *dielectric constant* whose value is different for different materials. Dielectric effects also detune the signal—that is, shift its frequency to a value that is not in resonance with the frequency for which the antenna is tuned.

Diffraction

Diffraction refers to the bending of an EM wave when it strikes the sharp edges or when it passes through narrow gaps. Due to diffraction, the receiver antenna will not receive the wave energy that it would have otherwise.

Free Space Loss

If the space through which the RF wave travels is free of all obstructing material and as a result there are no affects such as absorption, reflection, refraction, and scattering, there will still be some loss in signal strength, called *free space loss (FSL)*. This loss occurs simply due to the way a wave travels. An RF wave transmitted from a source travels in all directions in the form of an expanding sphere (called a *wavefront*), and therefore the power density (power per unit of surface area of this sphere) decreases as a result of this spreading out. If R is the distance from the transmitter antenna, the surface area of the sphere with radius *R* around the antenna is $4\pi R^2$. Therefore, the power density (and hence the signal strength) of a propagating wave at a point in space is inversely proportional to the square of distance of this point from the transmitter antenna. In other words, the free space loss will be directly proportional to the square of this distance. In addition, the loss is inversely proportional to the square of the wavelength of the propagating wave.

The FSL is measured using the following equation:

$FSL = (4\pi Rk/\lambda)^2$

$FSL (dB) = 10 \log (4\pi Rk/ \lambda)^2 = 10 \log (4\pi Rfk/ c)^2 = 20 \log (4\pi k/c) +$
$20 \log R + 20 \log$

$f =>$

$FSL (dB) = 20 \log R + 20 \log \lambda + K$

where:

$K = 20 \log (4\pi k/c)$ and *k* is a constant that depends on the communication link and the units used for distance and wavelength.

Exercise 2.6

Assume that the value of the constant *K* is –147.5 when the frequency is measured in Hz, *d* is measured in meters, and the isotropic antennas are used for the communication link. What is free space loss for a 950 MHz signal at 1000 meters from the transmitter antenna?

Solution:

$f = 950 \times 10^6$ 1/sec

d = 1000 m

K = –147.5

FSL (dB) = 20 log R + 20 log λ + K = 20 log (1000) + 20 log (950 x 10 6) - 147.5

= 60 + 2- (8.98) - 147.5 = 60 + 179.60 - 147.5 = 217.5 = 92.1

FSL = 92.1 dB

Interference

Interference is the interaction between two waves. The signal wave can interact with other waves that it meets on the way to its destination. A resultant wave is produced as a result of interference, and the receiver receives the resultant wave. The interference can be constructive, in which case the resultant wave has a larger amplitude, or destructive, in which case the resultant wave has a smaller amplitude than the original wave.

Reflection

Reflection is the abrupt change in direction of a wavefront at an interface between two dissimilar media so that the wavefront returns into the medium from which it hit the interface. Radio waves are reflected when they strike objects much larger than the wave, such as floor, ceiling, and support beam. Metals are obstructions to the signal because they are good at reflecting RFID waves.

Refraction

Refraction is the change in direction of a wavefront at an interface between two dissimilar media, but the wavefront does not return to the medium from which it hit the interface. In other words, the radio waves bend when they pass from one medium into another. Figure 2.10 illustrates reflection and refraction.

Figure 2.10 Reflection and Refraction

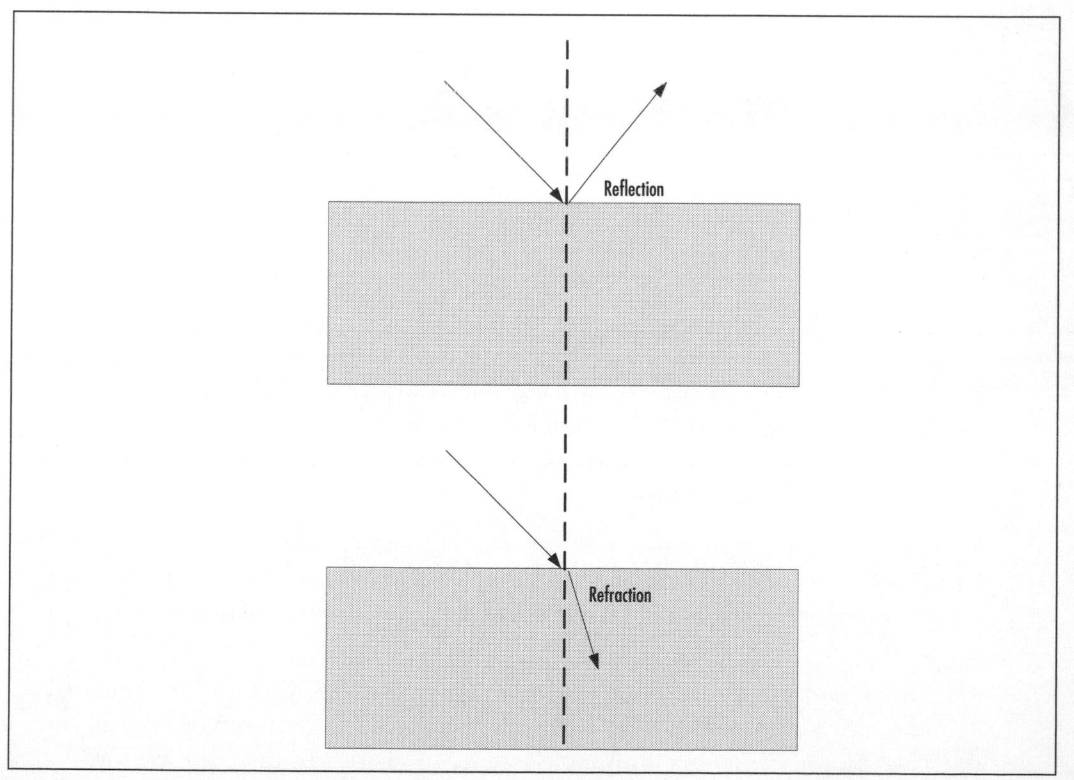

Scattering

Scattering is the phenomenon of absorbing a wave and reradiating it, thereby changing its direction. For example, reflection of an EM wave is actually a scattering. When a RF wave is scattered, it results in the loss of the signal or dispersion of the wave, as shown in Figure 2.11. It happens due to the interaction of the wave with the medium at the molecular level.

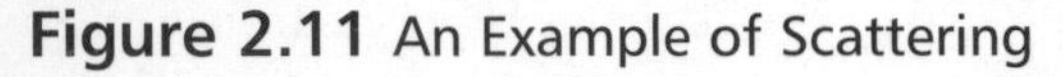

Figure 2.11 An Example of Scattering

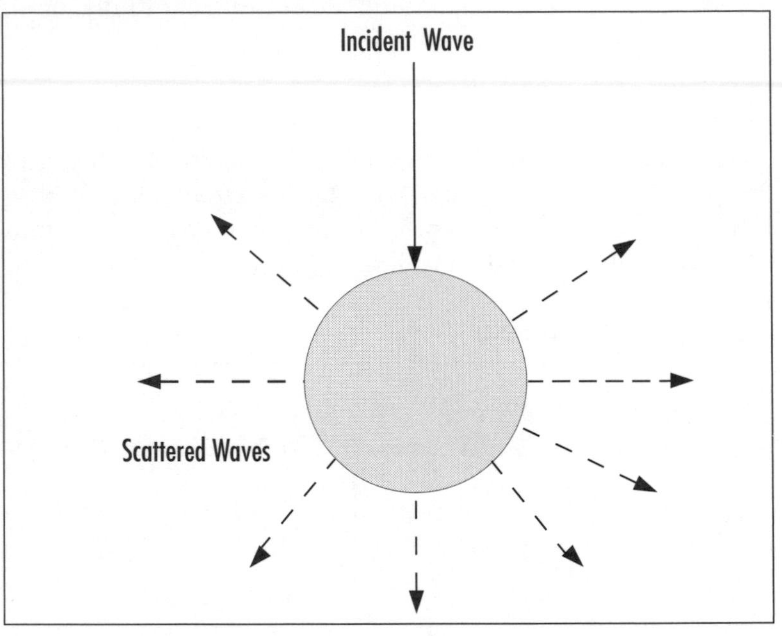

The three most important takeaways from this chapter are the following:

- The source encodes data (information) into the carrier signal using a modulation technique and sends it to the antenna through a transmission line. The input antenna impedance must match with the characteristic impedance of the transmission line to achieve optimal results.
- The antenna transmits the modulated carrier signal (carrying the information) into the free space. The polarizations and orientations of the transmitting and receiving antennas should be consistent with each other to maximize energy transfer.
- The hazards on the way from transmitting antenna to receiving antenna may affect the communication in a negative way. These hazards either weaken the wave by, for example, absorbing its energy or change its direction by, for example, reflecting it.

Summary

The readers and tags in an RFID system communicate with each other through RF waves. Encoding the data (to be communicated) into an RF wave (carrier signal), the emission of the RF wave by antennas, and the propagation of the RF waves between the antennas are governed by underlying physics principles. The data is encoded into the RF wave using modulation techniques. From performance viewpoint, there are two main factors in the RFID communication: the strength of the signal and the direction of the signal. In other words, you must understand all the characteristics that result in either the loss of power in the signal or the change of direction of the signal. For example, cable loss, impedance, and voltage standing wave ratio (VSWR) are important factors that affect how strong a signal antenna gets from the source through the transmission line. Because readers are directing their signal at the tags, the antennas used in RFID systems are directional antennas. Therefore, directivity, antenna gain, and polarization are important physical quantities that impact the performance of antennas.

Once the antenna radiates the RF waves into the free space, performance indicates how intact it will reach its destination. This part of the performance depends on factors such as absorption, reflection, refraction, and scattering. Water is a good absorber, and metals are good reflectors. RFID systems typically use two kinds of communication technique: inductive coupling to communicate within the near field and backscattering to communicate in the far field. Inductive coupling is used by RFID systems operating at LF and HF because the high wavelengths corresponding to these high frequencies will require ridiculously large antennas.

Most of the physics behind RFID relates to how readers and tags communicate with each other. In the next chapter, we discuss tags in greater detail.

Exam's Eye View

Comprehend

☑ Inductive coupling is the process of transferring energy from one circuit to another through a shared magnetic field by virtue of the mutual inductance between the two circuits.

☑ Backscattering is the process of collecting inbound signal (energy), changing the signal (the data it carries), and reflecting it back to where it came from.

- ☑ Inductive coupling is used by LF or HF RFID systems, whereas backscattering is used by long-range RFID systems operating at UHF or microwave frequencies.
- ☑ For a maximum transfer of power, the reader and the tag antennas should have the same polarization.
- ☑ If the receiving antenna is circularly polarized, it will receive some radiation regardless of the polarization of the transmitting antenna. This is because a circular polarization has both components of the linear polarization: horizontal and vertical.

Look out

- ☑ Inductive coupling works only in the near field of the RF signal.
- ☑ RFID systems operating at LF or HF use inductive couplings because it's not practical to use antennas for transmitting (or receiving) EM waves in this frequency range: low frequency means high wavelength, which requires very large antennas.
- ☑ If the transmitting antenna is horizontally polarized and the receiving antenna is vertically polarized (or vice versa), not much power transfer is going to happen.
- ☑ The UHF waves, due to their shorter wavelengths, are more susceptible to absorption than LF and HF waves.

Memorize

- ☑ In an environment of poor directivity, a reader may end up reading a tag outside its zone. This is called a phantom read or a ghost read.
- ☑ Water and the objects containing water such as liquid products, wood, and food are especially good at absorbing RF waves.
- ☑ Metals are good at reflecting RF waves.
- ☑ Diffraction is caused by sharp edges and narrow gaps such as slits.

Key Terms

Antenna The device used to transmit and receive signals such as radio waves. Both a reader and a tag have their own antennas through which they communicate with each other.

Antenna gain Ratio of energy radiated at a point of maximum radiation from an antenna to the energy radiated at the same point by some reference antenna.

Attenuation Decrease in the amount of something. In RF physics, it means the decrease in amplitude (strength) of the RF signal (wave).

Backscattering The process of collecting an inbound signal (energy), changing the signal (the data it carries), and reflecting it back to where it came from.

Beamwidth The angle between the two half-power points around the point (the main lobe) that has the peak effective radiated power.

Cable loss The amount of signal power lost in the cable being used as a transmission line.

Characteristic impedance The impedance of the transmission line when it's assumed to be lossless and of infinite length.

Carrier signal The wave that carries the data signal.

Data signal The wave that actually contains the information that needs to go to the receiver.

Diffraction The bending of an EM wave when it strikes sharp edges or when it passes through a narrow gap (slit).

Directivity The ability of an antenna to focus in a particular direction to transmit or receive energy. It is calculated as the ratio of the maximum value of power transmitted (or received) per unit of solid angle to the average power transmitted (or received) per unit of solid angle.

Effective radiated power The power that will need to be supplied to a reference antenna to produce the same power as this antenna is radiating in a specific direction.

Far field The EM radiations beyond the antenna's near field. In the far field, the signal power decreases as square of the distance from the antenna.

Impedance Resistance to the flow of current in a circuit element, measured as a ratio of voltage across the element and current through the element.

Interference The interaction between two waves. The signal wave can interact with other waves that it meets on the way to its destination. A resultant wave is produced as a result of interference, and the receiver receives the resultant wave.

Link margin Refers to the ratio of maximum effective signal strength received to the minimum signal strength received. In RFID, it means the amount of power that a tag can extract from the RF signal before the communication between the tag and the reader weakens.

Modulation The process that encodes the data signal into the carrier signal and creates the radio wave that the antenna actually transmits to propagate.

Near field The EM radiations within the distance of the order of one wavelength from the antenna. In the near field, the signal power decreases as a cube of the distance from the antenna.

Noise An unwanted electrical wave (or energy) present in a circuit or in a signal.

Polarization Refers to the direction of oscillations in the EM waves transmitted by the antenna.

Reflection The abrupt change in direction of a wave at an interface between two dissimilar media so that the wave returns into the medium from which it hit the interface.

Refraction The change in direction of a wave at an interface between two dissimilar media, but the wave does not return to the medium from which it hit the interface.

Resonance The characteristic of a system to absorb more energy when the frequency of its oscillations matches the system's natural frequency (resonant frequency) than it does at other frequencies.

Scattering The phenomenon of absorbing a wave and reradiating it, thereby changing its direction.

Standing wave A pattern of waves produced from the interference of two waves of the same frequency traveling in opposite directions on the same transmission line.

Voltage standing wave ratio (VSWR) The ratio of maximum voltage to minimum voltage along the transmission line.

Wavefront Refers to the geometrical shape of the space occupied by a traveling wave. For example, an EM wave from an isotropic antenna travels in the free space in all directions, making spherical wavefronts.

Self Test

A Quick Answer Key follows the Self Test questions. For complete answers and explanations to the Self Test questions in this chapter as well as the other chapters in this book, see **Appendix A**.

1. Which of the following is *not* a true statement about RF waves?
   A. RF waves are EM waves.
   B. The wavelength of an RF wave increases as the wavelength decreases.
   C. The wavelength of an RF wave increases as the amplitude is decreased.
   D. For a given frequency, the wavelength of an RF wave will be shorter if it travels through the transmission line compared to traveling through the free space.
2. The signal-to-noise ratio is measured in which of the following units?
   A. Volts
   B. Ohm
   C. Watts
   D. Decibels
3. Which characteristic of an RF wave causes it to bend when it travels from air through water?
   A. Reflection
   B. Refraction
   C. Diffraction
   D. Absorption
4. Which of the following is not the modulation technique used in RFID systems?
   A. Amplitude modulation
   B. Frequency modulation
   C. Amplitude shift keying
   D. Amplification
5. Which of the following is not the characteristic that affects the propagation of an RF wave though imperfect free space?

A. Reflection

B. Scattering

C. Voltage standing wave ratio (VSWR)

D. Diffraction

6. Which of the following are the two communication techniques used by RFID systems? (Choose two.)

A. Backscattering

B. Inductive coupling

C. Scattering

D. Link margin

7. An RF signal travels the fastest through:

A. Water

B. Free space (air)

C. Glass

D. Transmission line

8. Impedance mismatch will generate all the following effects except:

A. Interference

B. Standing wave

C. Reflection

D. Diffraction

9. All the following are true about a horizontally polarized wave except:

A. It travels parallel to the surface of the Earth.

B. Its electric field vector is parallel to the direction of the wave propagation.

C. Its magnetic field vector is perpendicular to the direction of the wave propagation.

D. It is generated by a linearly polarized antenna.

Self Test Quick Answer Key

For complete questions, answers, and explanations to the Self Test questions in this chapter as well as the other chapters in this book, see the Self Test Appendix.

1. C
2. D
3. B
4. D
5. C
6. A and B
7. B
8. D
9. B

PV27

Chapter 3

RFID+

Working with RFID Tags

Exam Objectives	What it Really Means
4.1 Classify tag types.	Understand the types (active and passive) and classes of the tags. You must understand that the passive tags can operate at all frequency ranges: LF, HF, and UHF. Especially don't forget the UHF passive tags. You must also understand that tags at LF and HF often use inductive coupling, and passive tags can use backscattering.
4.2 Given a scenario, select the optimal locations for an RFID tag to be placed on an item.	You must understand that the requirement in placing a tag is that it must be easily read by a reader. Also understand that tags need to be put into different forms before they can be attached to items. You must also know that different application requirements have given rise to different forms of tags such as inserts, smart labels, and tie-on tags. Understand the issues regarding the placement of tags such as tag orientation, material of and around the tag, and selecting the adhesive to attach the tag.

Introduction

The items that you need to identify and track are tagged with, well, tags. So, a tag is the "better half" of the RFID system because it contains information about the item to which it is attached and has the capability to provide that information on request. A tag makes it to the item in three steps: The tag with the basic functionality is manufactured, the tag is turned into a label, and the label is placed on the item. From your perspective, this process involves the following facts:

- All the tags are composed of the same basic components because they offer the same basic functionality: to help identify and track an item.
- To meet the varied needs of different applications, tags come in different forms, shapes, and sizes.
- Tags must be properly placed on items so that they could be easily read by readers.

So, the main goal of this chapter is to understand the role of a tag in an RFID system. To accomplish this goal, we will explore three avenues: tag components, tag types, and tag placement.

Understanding Tags

Generally speaking, RFID is a means to identify an object using radio frequency transmission, which suggests that communication is involved in the identification process. The communication takes place between a reader and a tag. A tag, attached to an item that needs to be tracked, contains identification and possibly more information about the item. For example, in a supply-chain system, a tag may contain the following information about an item: source, destination, and route.

You need to know what makes a tag—that is, its components—and what it looks like, including its size and shape.

Components of a Tag

The components of a tag are there to support its functionality by:

- Storing the information about an item
- Processing the request for information coming in from a reader
- Preparing and sending the response to the request

To support this functionality, a tag, as shown in Figure 3.1, consists of the following three main components:

- **Chip** The chip is used to generate or process a signal. It's an integrated circuit (IC) made of silicon. The chip consists of the following functional components:
  - **Logical unit** Implements the communication protocol used for tag-reader communication.
  - **Memory** Used to store data (information).
  - **Modulator** Used for modulating the outgoing signals and demodulating the incoming signals.
  - **Power controller** Converts the AC power from the incoming signal to DC power and supplies power to the components of the chip.

  The chip is connected to the antenna so that it can send the outbound signal to the antenna and can receive the inbound signal from the antenna.

- **Antenna** In an RFID system, a tag's antenna receives the signal (a request for information) from a reader and transmits a response signal (identification information) back to the reader. It's made of metal or a metal-based material. Both readers and tags have their own antennas. You learned about antennas in Chapters 1 and 2, and you will learn more details about them in Chapter 6. In this chapter, it is sufficient to know that a tag's antenna radiates and receives radio waves to transmit and receive a radio signal. Furthermore, note the following two points:
  - The antennas are usually used by tags (and readers as well) operating at UHF and microwave frequencies.
  - The tags (and readers) operating at LF and HF use inductive coils (as antennas) to send and receive signals in the inductive coupling communication technique. As you know from Chapter 2, the size of a traditional antenna for sending or receiving an LF signal would need to be ridiculously high due to the high wavelengths of these signals.

  Both the chip and antenna are housed on a substrate.

CAUTION

Note two important points: Both readers and tags have antennas, and a tag (and a reader) that uses inductive coupling as its communication technique uses an inductive coil for an antenna instead of a standard antenna.

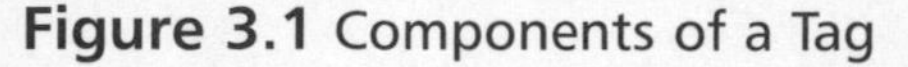

Figure 3.1 Components of a Tag

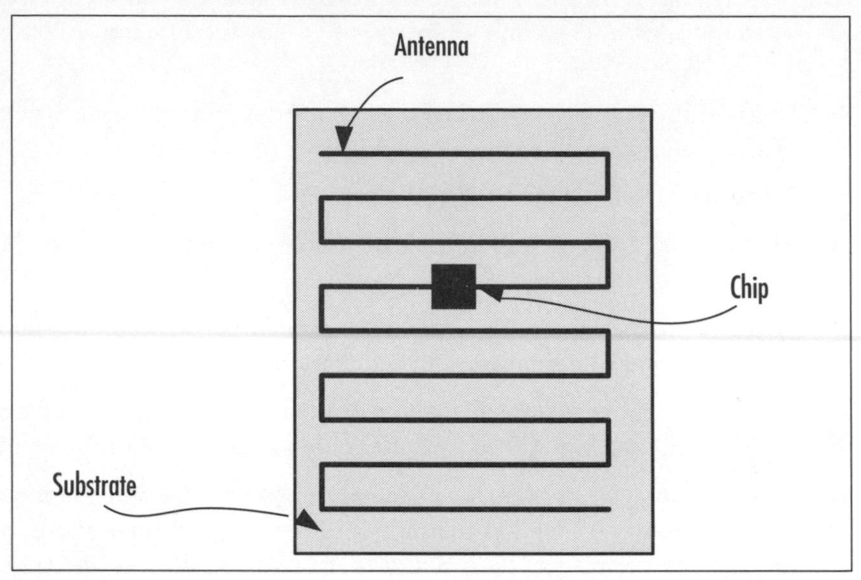

- **Substrate** This is the layer that houses the chip and the antenna. In other words, it's the support structure for the tag. Substrates can be made of different materials such as plastic, polyethylene terephthalate (PET), paper, and glass epoxy. Substrate material can be rigid or flexible, depending on the usage requirements.

  Substrates for RFID tags are designed to meet specific usage requirements such as the following:

  - Dissipation of static charge buildup
  - Durability under specific operating conditions
  - Mechanical protection for chip, antenna, and connections
  - Smooth printing surface

So, a tag consists of a chip and an antenna housed on a substrate. Now this thing is going to be attached to an item, so a natural question to ask is: How big is a tag? In other words, depending on the item to which a tag will be applied, tag size matters.

Tag Size

The preferred tag size might depend on the item on which the tag will be applied and the environment in which the item exists. To meet the varied requirements of different applications, tags come in various shapes and sizes. Here are some examples:

- Large tags that are several inches in length, width, and height can be used to track large objects such as vehicles like trucks and rail cars.
- Rectangular shaped tags can be used as antitheft devices.
- Thin tags can be applied under a paper or plastic label on individual items such as books or packages such as boxes.
- Screw-shaped tags can mark and track specific trees.
- Inserting tags the size of a pencil lead (less than half an inch in length) under the skin can help track animals.

The smallness of a tag is limited by the antenna size. To select the right tag for a given environment, you must understand the tag types and operating frequencies.

Operating Tag Frequencies

To respond to readers, tags use radio waves, which are basically the electromagnetic waves covering part of the electromagnetic spectrum of frequencies called *radio frequency spectrum*. Because the RFID systems generate and radiate the electromagnetic waves that fall in the radio frequency spectrum, they are justifiably classified as radio systems. However, other radio services have been operating before the arrival of RFID systems. Radio, television, mobile radio services (police, security services, and industry), marine and aeronautical radio services, and mobile telephones are a few. Therefore, it is important to ensure that these services are not disrupted or impaired by the RFID newcomers. This requirement significantly restricts the suitable operating frequency ranges available for RFID systems. Therefore, the so-called industrial, scientific, and medical (ISM) frequencies, originally reserved for noncommercial uses in industrial, scientific, and medical fields, are generally used for RFID systems.

Table 3.1 shows the radio frequency ranges that are of interest to RFID systems, along with the ISM frequencies. RFID systems use many different frequencies in the radio frequency spectrum, but there are four most commonly used radio frequency ranges: low frequency (30–300 KHz), high frequency (3–30 MHz), ultrahigh frequency (300 MHz–3 GHz), and microwave frequencies (1 GHz–300 GHz).

Table 3.1 Radio Frequency Ranges in Which RFID Systems Can Operate and the Corresponding Read Ranges for Passive Tags

Name	Frequency Range	Wavelength Range	ISM Frequencies	Read Range for Passive Tags
Low frequency (LF)	30–300 kHz	10 km–1 km	<135 kHz	<50 cm
High frequency (HF)	3–30 MHz	100 m–10 m	6.78 MHz, 8.11 MHz, 13.56 MHz, 27.12 MHz	<3 m
Ultrahigh frequency (UHF)	300 MHz–3GHz	1 m–10 cm	433 MHz, 869 MHz, 915 MHz	<9 m
Microwave frequency	3 GHz–300 GHz	30 cm–1 mm	2.44 GHz, 5.80 GHz	>10 m

Table 3.1 also shows the read ranges for a passive tag (a tag that does not have its own source of power, such as battery) corresponding to each frequency range. An active tag (a tag that has a battery) can have a read range of up to 100 meters. For example, active tags used on large assets such as cargo containers, rail cars, and large reusable containers, which usually operate at 455 MHz, 2.45 GHz, or 5.8 GHz, typically have a read range of 20 meters to 100 meters.

Note that the read range performance improves with the increase in the frequency. However, for a given frequency, the read range also depends on other factors, such as the maximum power the antenna is allowed to transmit, the communication technique being used, and the tag type.

CAUTION

For a given frequency, tag type, and communication technique, the practical read distance of a tag also depends on other factors such as the regulated maximum radiated power and antenna size.

RFID systems operating in the LF and HF ranges typically use the same frequencies all over the world, as shown in Table 3.1, but there is no global agreement on which frequencies should be used for RFID systems operating in the UHF range. As shown in Table 3.2, different UFH frequency bands are allocated to the RFID systems in different regions of the world.

Table 3.2 UHF Frequency Bands Allocated for the RFID Systems Around the Globe

Area	UHF Frequency Band Allocated to RFID Systems	Power
United States	902–928 MHz	4 W
Australia	918–926 MHz	1 W
Europe	865–868 MHz	2 W
Hong Kong	865–868 MHz 920–925 MHz	2 W 4 W
Japan	952–954 MHz	4 W

This section mentioned active tags and passive tags. What are they? Let's take a look.

Understanding Tag Types

The two major characteristics that determine the performance and use of a tag are the tag type and the frequency at which the tag operates. The tag types are determined by the following two factors:

- Can the tag initiate the communication?
- Does the tag have its own power source?

Based on different combinations of answers to these two questions, there are three types of tags: passive, semipassive, and active.

Passive Tags

A *passive tag* is a tag that does not have its own power source, such as a battery, and therefore cannot initiate the communication. It responds to the signal sent by the reader by taking power from the reader's signal. In other words, the reader's signal wakes up the passive tag. Here is how it works:

1. The passive tag's antenna (or coil) receives the signal from the reader.
2. The antenna sends the signal to the IC.
3. Part of the signal power is used to power up the IC.
4. The IC powers up, processes the incoming signal, and sends the response.

The characteristics of a passive tag include the following:

- **Placement** Because a passive tag entirely depends on the reader for its power, it must be inside the interrogation zone to get enough power to generate a response.
- **Size and range** Because there is no battery, passive tags tend to be smaller in size and have a shorter read range compared to active tags.
- **Lifespan** Because there is no need to replace a battery, passive tags have a longer life.
- **Memory** The memory capacity of passive tags varies from 1 bit to several kilobytes.

Remember the following about passive tags:

- Passive tags can operate at any of these frequency ranges: LF, HF, and UHF.
- Depending on the frequency range at which a passive tag is operating, a passive tag may have a read range from 2 millimeters to about 5 meters.
- The passive tags are simpler and cheaper and therefore more popular.
- LF passive tags are ideal for applications that require reading from a close range.

CAUTION

It's very easy for a beginner to fall into the trap of thinking that passive tags operate only at low frequencies and have a very short read range. At least from the exam's perspective, you should burn a term into your head: *UHF passive tags*.

So, the defining characteristic of a passive tag is that it does not initiate communication. If a tag does have a battery but does not initiate communication, it is still a passive tag, but it's called a *semipassive tag*.

Semipassive Tags

A *semipassive tag* is a tag that has its own power source such as a battery but does not initiate communication. It responds to the signal sent by the reader by taking power from the reader's signal. In other words, the reader's signal wakes up the passive tag. A passive tag uses its battery to run its circuitry. The characteristics of a semipassive tag include the following:

- **Operation** Because a semipassive tag can transmit a response signal only if it gets adequate power from the reader, its operating principle is very similar to that of a passive tag.
- **Size and range** Because a semipassive tag has its own battery, it is larger than the passive tag. For the same reason, it can produce a stronger signal, which can transmit across a longer distance, resulting in a larger read range compared to a passive tag.
- **Lifespan** A semipassive tag has a shorter life (tied to battery) than a passive tag.
- **Memory** The memory capacity of a semipassive tag varies and can be greater than that of a passive tag, partly due to its larger size (more room for components) and battery.

In a nutshell, a semipassive tag uses a battery to run the circuitry but still does not initiate communication because it still uses the power from the incoming signal to prepare the response.

So, on one end of the spectrum is the passive tag that contains no battery and cannot initiate communication. In the middle is the semipassive tag that has a battery but does not initiate communication. On the other end of the spectrum is the tag type that contains the battery and can initiate communication; this tag is called an *active tag*.

Active Tags

An *active tag* is a tag that has its own power source such as a battery and can initiate communication by sending its own signal. It does not rely on the power from the reader to run its circuitry or to create the signal. It does not need a wakeup call from the reader. The characteristics of an active tag include the following:

- **Operation** Because an active tags has its own power source, it has the choice of staying up all the time or waking up when a signal is received. A tag that is operating all the time can broadcast its location at predetermined intervals.
- **Size** Because of their power sources (batteries), active tags are the largest in size. Typical sizes are (1.5 x 3) x 0.5 $inch^3$. However, with the advancement of technology, the smallest active tags could be the size of a coin.
- **Read range** Because an active tag has its own power source for circuitry and for generating signals, it can achieve the greatest read range. Some active tags have the ability to send a signal across a distance of 1 km. However, confined to standards and regulations, many active tags have read ranges of tens of meters. Due to its larger read range, an active tag can be integrated with a global positioning system (GPS) to pinpoint the exact location of an object.

- **Lifespan** Finite but long enough battery lifetime. It can be as long as 10 years.
- **Memory** The memory capacity of passive tags varies and can be greater than that of passive and semipassive tags, partly due their larger size (more room for components) and batteries.

Active tags can be used to track high-value assets such as rail cars and cargo containers that need to be read from large distances. Because active tags have the ability to initiate communication, they can be further divided into two subtypes:

- **Active transponders** These tags are activated only when they receive a signal from a reader. This way the tags prolong their battery life. These tags can be used in applications such as toll collection systems and checkpoint control systems.
- **Beacons** A *beacon* is a tag that emits a signal at predetermined intervals. Beacons are mostly used in real-time locating systems (RTLS). Possible applications for beacons include the following:
    - Tracking parts in large manufacturing facilities
    - Marine and aircraft rescue operations

NOTE

Active tags usually operate in the UHF and microwave frequency ranges (455 MHz, 2.45 GHz, and 5.8 GHz) and have read ranges from 20 to 100 meters.

EXERCISE 3.1

Describe how an active tag can work in a toll collection system.

Solution: An active tag is attached to a car. As the car approaches a toll booth, a reader installed at the booth sends out a signal to the approaching tag. The tag responds by sending back a signal that contains the tag ID and possibly other information.

EXERCISE 3.2

Describe how a beacon can be used in a ship rescue operation.

Solution: A beacon is attached to a ship and is emitting signals at preset intervals. These signals contain the information about the ship. Using this information, satellites or other sensor systems can track down the ship.

The characteristics of passive, semipassive, and active tags are summed up in Table 3.3.

Table 3.3 Characteristics of Tag Types

	Tag Type		
Tag Characteristic	**Passive**	**Semipassive**	**Active**
Power source	No power of its own; receives power from the reader's signal	Has its own power source (battery)	Has its own power source (battery)
Communication	Communication must be initiated by the reader	Communication must be initiated by the reader	Can respond to the reader's signal and can also initiate the communication
Size	Small Could be as small as (0.15 mm x 0.15 mm) x 7.5 µm	Medium	Largest, typically (1.5 x 3) x 0.5 $inch^3$
Read range	Short 2 mm; few meters depending on the operating frequency	Up to 100 m	Large (up to 1 Km is possible); some limitations apply, resulting from standards and regulations
Memory design	Read only (RO), write once/read many (WORM), or read/write (RW)	Read only (RO), write once/read many (WORM), or read/write (RW)	Read only (RO), write once/read many (WORM), or read/write (RW)
Memory capacity	Mostly up to 128 bits, but some tags can have memory up to 64 KB	—	Up to 8 MB
Cost	Inexpensive	Intermediate	Expensive

NOTE

Passive tags are simple and less expensive. UHF provides the greater read range. To get the best of the both worlds, companies are increasingly becoming interested in using UHF passive tags, especially in the supply chain. For example, consider a warehouse in which a reader must be able to read a tag from about 3 meters distance. The LF and HF tags would need to be read from much closer distances; therefore, the reader begins to interfere with the normal operation of equipment such as forklifts.

So, the tag types categorize the tags from the communication perspective—that is, whether a tag can initiate communication and whether the tag has its own power to generate the communication signal. As tag technology progresses, other ways of categorizing tag types are developing. One of them is categorization by class, which is largely based on memory design.

Tag Classification

As you know by now, tags are data holders that are attached or affixed to an object and carry that object's data, including its identification number. These numbers are also called *electronic product codes (EPCs)*, and the tags containing them are called *EPC tags*. The complexity of an EPC tag varies depending on its functionality—how it communicates and whether or not it has a power source of its own.

NOTE

The EPC is a group of coding schemes for tags defined by the standard called Generation 2. These coding schemes are designed to meet the wide spectrum of needs of various industries while guaranteeing uniqueness of codes for all tags that comply with the standard. The EPC was originally the creation of the Massachusetts Institute of Technology (MIT) Auto-ID Center, a consortium of over 120 corporations and university research labs.

The increased functionality and therefore complexity of tags results in increased cost because tags with advanced functions require more expensive microchips and their own power source. Although most business sectors require only the simplest and therefore lowest-cost tags, the potential value of complex tags justifies their increased cost in certain industries. For example, think of the food industry, which might want to add temperature tracking by adding a temperature sensor on tags attached to food items in containers. To accommodate varying levels of complexity, MIT's Auto-ID Center proposed six tag classes, presented in Figure 3.2. As illustrated in the figure, each class is a subset of the functionality contained within the higher class.

Figure 3.2 Classes of Tags

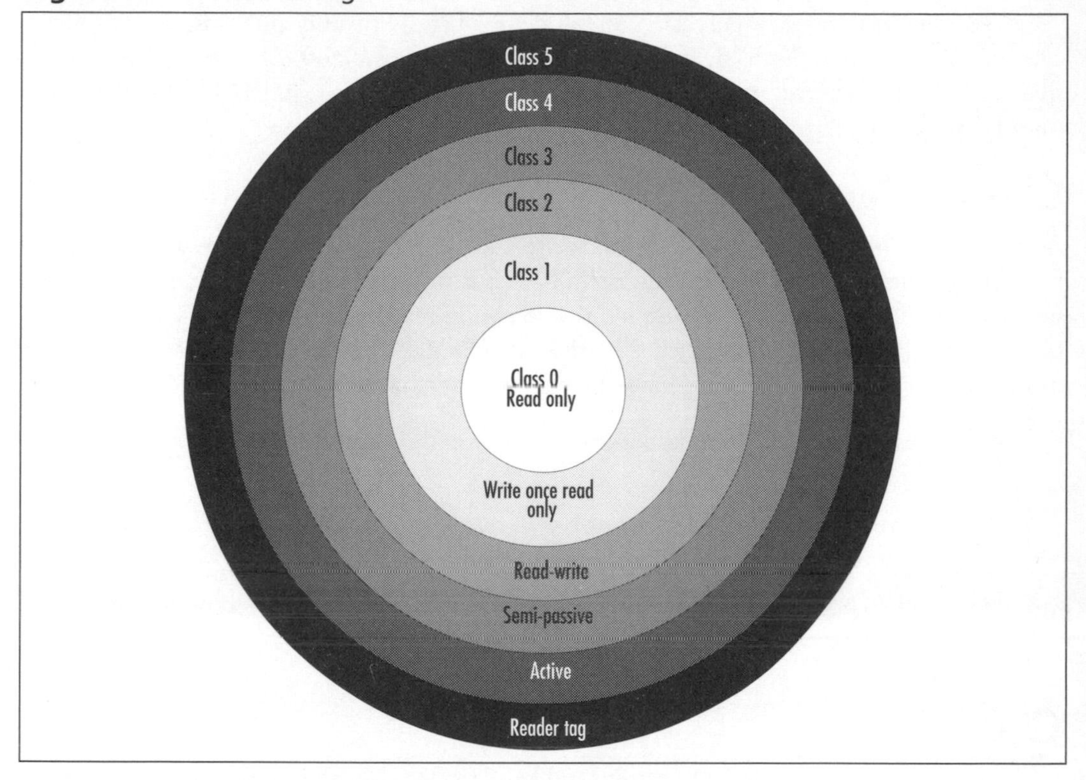

These tag classes are discussed in the following sections.

Class 0 Tags

A *class 0 tag* is a simple, passive, and read-only tag that is programmed with a unique EPC number during manufacturing. These tags cannot be programmed by users in the field. Being UFS based and low cost, these tags are suitable for applications that need to detect only the tag's presence and not any other data, such as antitheft devices. A class 0 tag must have the following elements:

- **The tag identifier (TID)** This is assigned by the manufacturer to uniquely identify the product.

- **EPC number** This number is assigned by the manufacturer to identify the specific object to which the tag is attached. This is also called an *object ID (OID)*.
- **A kill function** This can be used to disable the tag permanently.

Class 0+ tags are essentially the same as class 0 tags, with only one difference: You can write to class 0+ in the field, but only once. Both class 0 and class 0+ tags are passive tags—that is, they do not have a power source of their own, and as a result they can only be woken up by a signal from the reader.

Class 1 Tags

A *class 1 tag* is a simple, passive, read and write-once, backscatter tag. It has a one-time, field-programmable, nonvolatile memory. The tag is manufactured with no data written into the memory. However, because it's a write once/read-only memory (WORM) tag, the data can be written (once) either by the manufacturer before shipping or by the user in the field. The tags in this class have the following characteristics:

- **Passive** Cannot initiate communication and do not have their own power source.
- **Memory** Have 128-bit memory: 96 bits for storing identification and 32 bits for error correction and kill function.
- **Write-once, read-only memory (WORM)** Can be programmed by the manufacturer or by the user in the field, but only once.

Class 2 Tags

A *class 2 tag* is also a passive backscatter tag like class 0 and class 1 tags, but it's very flexible when it comes to memory. It has the following characteristics:

- **Passive** Cannot initiate communication and does not have its own power source.
- **Memory** Up to 65 KB of read/write memory.
- **Authenticated access control**

Class 2 tags are typically used to log data and therefore contain more memory than just what's needed to store identification. Class 1 and 2 tags have become popular among the majority of RFID applications. However, for certain applications with diverse requirements, you need class 3 tags.

Class 3 Tags

A *class 3 tag* is a semipassive backscatter tag that has onboard sensors. It has the following characteristics:

- **Semipassive** Cannot initiate communication but has its own power source to generate signals. Remains passive until activated by a reader signal.
- **Memory** Up to 65 KB of read/write memory.
- **Integrated sensor circuitry**

A class 3 tag with a built-in battery supports increased read range. The built-in memory also supports sensor recording parameters such as temperature, pressure, and motion into the memory without the power from the reader's signal. These tags can be used in supply chain applications such as on pallets and containers, to provide historical information.

Class 0, 1, and 2 tags are passive tags, whereas class 3 tags are semipassive. That means none of these tags can initiate communication, and they all take power from the reader's signal to generate a response signal. The needs of some applications may require the tag to use its own power source to generate signals and to be able to initiate communication. These requirements are met by class 4 and 5 tags.

Class 4 Tags

A *class 4 tag* is an active tag with integrated transmitter. It uses a built-in battery to run the microchip's circuitry and to power the transmitter to broadcast the signal to a reader. It has the following characteristics:

- **Active** It can initiate communication because it has its own power source to run the circuitry and to generate the signal.
- **Memory** Rewritable.
- **Communication** Ability to communicate with other tags.
- **Networking** Ad hoc networking capabilities.

The class 4 tags can be used in applications such as parents keeping track of their children in an amusement park, a tag inside a cargo container passing information from other tags to an external reader (networking), and tags working with GPSs to track objects globally.

Class 5 Tags

A *class 5 tag* is an active RFID tag that has the capability of communicating with other class 5 tags and other devices. Its capabilities include all the capabilities of a class 4 tag. The only additional functionality that a class 5 tag has over a class 4 tag is its ability to initiate communication with all classes of tags. That means it can initiate communication with a passive tag as well by waking up the tag. Due to this reader functionality, a class 5 tag is also called a reader tag.

All these classes of tags are summarized in Table 3.4.

Table 3.4 Characteristics of RFID Tag Classes

	Tag Characteristic			
Tag Class ||v	**Type**	**Memory**	**Communication**	**More Properties**
Class 0	Passive	Read-only	Does not initiate communication	The EPC number is encoded onto the tag during manufacture and can be read by a reader)
Class 0+	Passive	Same as class 0, but you can write once	Does not initiate communication	—
Class 1	Passive	Read and write-once	Does not initiate communication	EPC number is not encoded by the manufacturer but can be encoded later in the field
Class 2	Passive	Read and write-once	Does not initiate communication	Encryption
Class 3	Semipassive	Read and rewritable	Does not initiate communication	Class 2 capabilities plus extra such as integrated sensors
Class 4	Active	Read and rewritable	Can initiate communication; power their own communication; tag-to-tag communication possible	Class 3 capabilities plus extras
Class 5	Active	Read and rewritable	Can initiate communication; power their own communication; tag-to-tag communication possible	Class 4 capabilities plus extras

So, tags are classified to be manufactured with varied levels of features, capabilities, and resulting complexity. As Table 3.4 depicts, each higher-class tag offers all the features and capabilities of the lower-class tags and more.

A common core functionality of all kinds of tag is that they can be read by readers. The maximum distance from which a tag can be read is called its *read range*. We need to take a close look at tags' read ranges.

Read Ranges of Tags

The need for read range of a tag generally corresponds to the application requirement. In other words, the read range of a tag plays an important role in determining which application will use this tag. For example, in a warehouse, a reader must be able to read tags from a distance of a few meters, say about 3 meters, at least. Otherwise, it will start interfering with the normal operation of equipment such as forklifts.

From the physics perspective, where does the read range come from? As shown in Figure 3.3, it is mainly determined by the following four characteristics:

- Operating frequency
- The maximum allowed power emission
- Tag type: active or passive
- Communication technique: inductive coupling or backscattering

Figure 3.3 Characteristics That Affect the Read Range: Operating Frequency, Regulated Power Emission, Communication Technique, and Tag Type

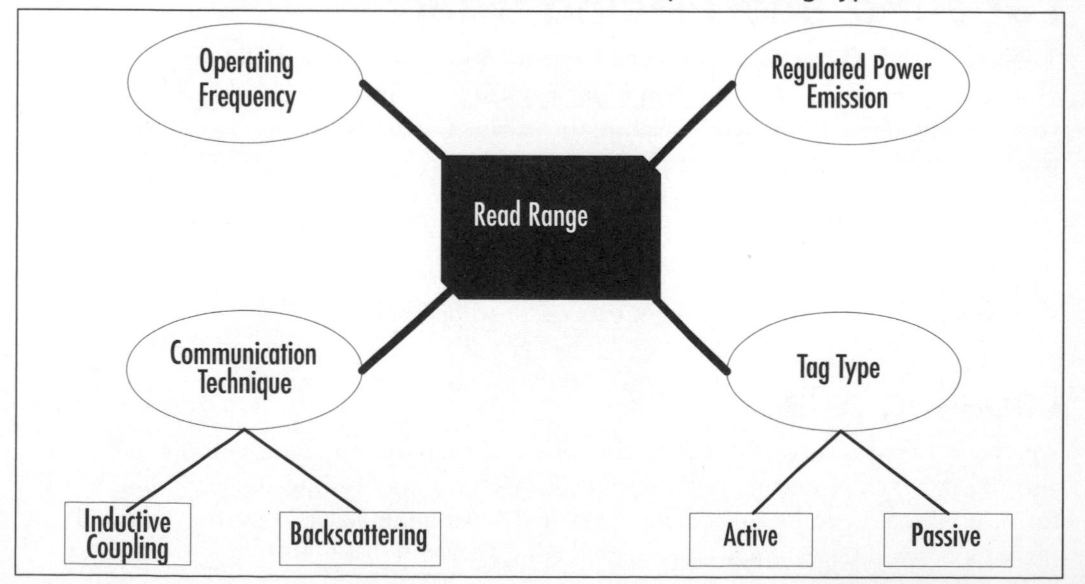

You have already seen in this chapter how the read range depends on the tag type and the operating frequency. The maximum allowed power that an antenna can emit comes from standards and regulations. The higher the power, the larger the read range. The read range also depends on which of the following two communication techniques your RFID system is using:

- **Inductive coupling** You learned about inductive coupling in Chapter 2. The reader and the tag use coils as antennas. These coils create magnetic fields. The variations in the magnetic field are used to transfer power (and data) between the reader and the tag. In technical terms, the energy is transferred between two circuits (tag and reader) by virtue of the mutual inductance between the circuits. This technique limits the read range because it only works in the near field of the coils. Therefore, inductive coupling requires that the reader be close to the tag. This leads to a read range of about 30 cm for LF tags to 1 meter for HF tags.
- **Backscattering** Also called *backscatter coupling*, this concept was also discussed in Chapter 2. Backscattering is typically used by UHF passive tags. Because backscattering works beyond the near field, it allows larger read ranges. For example, the read range of a UHF passive tag using backscattering can be larger than 3 meters.

So far, we have discussed tags from the perspectives of the functionality and the features offered by them. But before you can use these tags, they need to be changed to some kind of labels and placed on the items that that need to be identified and tracked.

Labeling and Placing a Tag

Tags are placed on the items that need to be identified and tracked. Before you can place a tag, that is, attach a tag to an item, it needs to be in a form so that it can be conveniently attached to the item. Creating that form is called *labeling the tag*. In other words, to tag an item is a three-step process:

- Manufacture the tag (a combination of IC, antenna, and substrate).
- Label the tag—that is, convert the tag into some kind of label.
- Place the label on the item that needs to be tracked.

Labeling a Tag

The basic functionality of all tags is the same, and therefore the basic components are also the same: chip, antenna, and substrate. However, a tag needs to be converted into a form in which it can be conveniently attached to an item. What form (e.g., size and shape) a tag will take depends on the following factors:

- The characteristics, such as the material of the item to which the tag is to be attached
- The environment around the item

Therefore, labeling a tag depends on its application, such as where and how it will be placed. Accordingly, tags are manufactured and turned into usable forms in various shapes and sizes. In other words, RFID tags are available in various media configurations. Your selection depends on the application requirements. This has given rise to different kinds of tags and terms that you should be aware of. Some of them are discussed in the following sections.

Inlay

The *inlay* is the bare-bones tag discussed in this chapter so far—that is, the combination of antenna, chip, and substrate. The inlay needs to be packaged (labeled) before use.

Insert

An *insert* is an inlay inserted between a label in the front and an adhesive layer in the back. The adhesive in the back can be permanently mounted, for example, on the inner wall of a tire. RFID inserts are available in different sizes depending on the applications in which they will be used. Here are some examples:

- Thick inserts intended to be used in harsh environments
- Paper-thin inserts in a pressure-sensitive environment, used, for example, to track parcels
- Postage stamp-sized inserts applied to videocassettes

Smart Labels

A *smart label* is a barcode label that has an embedded RFID tag inside it. You can print human-readable, useful information on the label face, such as sender's address, destination address, and product information. A smart label has the following components:

- **Inlay** IC, antenna, and substrate
- **Label face stock** Covers the top of the inlay and provides area for printing human-readable information.
- **Adhesive** Used to attach the face stock to the inlay.
- **Release liner** Covers the bottom of the inlay. This layer can be used to convert the pressure-sensitive inlay into rolls for easy and safe distribution. This layer can later be removed to place the smart label on an item.

Smart labels are designed to withstand a number of hazards such as extreme temperature, chemicals, moisture, and exposure to ultraviolet radiation.

Pressure-Sensitive Labels

Pressure-sensitive labels are used in RFID-enabled media and are basically the same as smart labels. A tag can also be inserted into an envelope, which can be attached to the face stock and coated with an adhesive for placement.

RFID-Enabled Tickets

RFID-enabled tickets are inserted into paper or plastic envelopes that are directly attached to the items that need to be identified. The paper or plastic material used for these envelopes should be ultraviolet-resistant—that is, transparent to UHF waves. Several types of polypropylene, polyester, polyethylene, and polyethylene terephthalate (PET) films are good selections and offer low attenuation for UHF waves.

Tie-On Tags

Tie-on tags are basically RFID-enabled tickets that are attached with a tie-on. These tags are usually used on nonconveyable items. However, you must consider that a tie-on made of a conductive wire may positively or negatively affect tag performance.

TIP

Avoid polyvinyl chloride (PVC) films with tags that use copper antennas because they run the risk of long-term antenna corrosion.

So, generally speaking, a label is any kind of tag attached to an item by an adhesive with the purpose of identifying the item. Labels come in many forms and can be differentiated by the type of base material, called *stock*, on which you can print, and by the *adhesive type* that they use.

Selecting Adhesive Types for Tags

An adhesive, used to affix a tag to the surface of the item that needs to be identified, is a strong chemical mixture. It may affect the tag's performance in various ways. For example, it might absorb RF waves and can corrode the antennas in the long run. Furthermore, some adhesives may deteriorate over time, even if they are perfectly fine at the time of their initial application. There are two basic types of adhesive:

- **Acrylic adhesives** Offer best high-temperature performance and the widest spectrum of properties.

- **Rubber-based adhesives** Offer high initial tack, good for applications such as tagging corrugated cases; relatively lower in cost.

The effectiveness of a tag adhesive depends on the following:

- Ease and strength of initial tack
- Time it takes the bond to form the full strength
- The final bond strength
- Long-term bond stability and resistance to deterioration over time
- Stability in a hostile environment

After a tag has been labeled, it can be placed on the item that needs to be identified and tracked.

Placing a Tag

Placing tags on the items can be challenging. There are no global guidelines regarding where on the items the tags should be placed. It depends on the application, item, and environment. On the other hand, you have only so much freedom in where you can place the tag on an item. Most of the time, the tag is placed on the outside of a package. In this case, you should be aware of how the package material can affect the RF signal. The material of the package or in the vicinity of the package can affect the RF signal between the tag and the reader in the ways shown in Table 3.5.

Table 3.5 Effects of Materials on RF Signals

Material	Effects on RF Signal
Corrugated cardboard	Absorption (caused by moisture soaked into the cardboard)
Conductive liquid	Absorption
Glass	Attenuation
Group of cans	Reflection, multiple paths
Human/animal body	Absorption, detuning, reflection
Metal	Reflection
Plastic	Detuning due to dielectric effect

NOTE

Corrugated means rippled at regular intervals. Usually shipping boxes are made with several layers of cardboard, and in the middle is a layer of corrugated paper or cardboard. This layer in the middle gives the box more cushion to withstand the hard knocks of shipping and handling.

The performance of an RFID system also depends on how the tagged items are packed together, for example, on a pallet. Following are some considerations.

Shadowing

Shadowing is caused when an item, say *B*, is behind another item, say *A*, from the perspective of the reader. In this case, a signal from the reader will be received only by *A*. The item *B* will get little or no signal and will be missed by the reader. Item *B*, in this situation, is said to be *shadowed* by item *A*. If the items inside a package or the cases on a pallet are densely packed, the reader will miss some of them.

NOTE

A *pallet* is a portable platform for stowing, handling, and moving cargo.

Tag Placement and Orientation

The orientation of a tag with respect to the reader is an important factor that impacts reading performance of an RFID system. For example, if a tag is oriented parallel to the direction of propagation of energy coming from the reader, it will not receive its signal, and no communication will occur. To facilitate communication, the tag antenna must face the reader antenna. For example, consider the following scenario.

Reader antennas are often mounted on gantries placed around a conveyor. The tagged packages (containers) are often cubic in shape, and therefore they have six faces to be considered for placing tags. In general, the bottom face should be avoided, to prevent mechanical damage to the tag. The top face should be avoided if there is a possibility that the packages will be stacked. Reader antennas on each side of the gantry will cover four faces of the container. The exact positions of the tag and the reader and the antenna orientations should be determined to ensure that the tag will be in the read zone of at least one reader. A single reader antenna may not see all the faces of the container, especially if the container contains RF-sensitive materials such as metals or liquids.

To improve performance, you can position antennas at different angles to read tags that would otherwise be missed.

The orientation of a tag also depends on the polarization of the reader's antenna.

Polarization and Orientation

As you learned in Chapter 2, antennas are either linearly polarized or circularly polarized. For optimal power transfer between the reader antenna and the tag antenna, the polarization of both antennas should match. The orientation of a tag's antenna should be consistent with the polarization of the reader's antenna. Here are some examples:

- In the case of a single dipole antenna, the antenna must be aligned parallel to the incoming field to receive it.
- If reader's antenna is producing horizontally polarized waves, the tag's antenna must be horizontally aligned.
- If the reader's antenna is circularly polarized, the tag will get some signal in any orientation because circularly polarized waves have both components, horizontal and vertical. In other words, circularly polarized reader antennas improve read performance.

TIP

When you do not know the tag orientation or you have no control over it, you should use the circularly polarized reader antenna because it can read horizontal tags, vertical tags, and tags aligned to angles between horizontal and vertical.

Our tag placement discussion so far mostly applies to readers and tags with antennas. How about the RFID system that uses inductive coupling for communication?

Orientation in Inductive Coupling

Readers and antennas that use the inductive coupling communication technique use inductive coils instead of antennas. These systems are relatively less sensitive to orientation. However, you should know that to transfer the maximum power, the two coils should be in the same plane. If one of them is rotated with respect to the other, the coupling (and therefore the power transfer) reduces in proportion to the cosine of the angle of rotation.

The three most important takeaways from this chapter are the following:

- The basic functionality of any tag is to store information about the item to which it is attached and to provide this information when requested by a reader. To support this functionality, all tags have three basic components: a chip, an antenna, and a substrate.
- There are two basic types of tags: active tags that can initiate communication and passive tags that cannot initiate communication.
- The principal consideration for placing a tag on an item is that it should be readable by a reader. A tag is useless if it cannot be read.

Summary

A tag is a component of an RFID system that contains the information about the item to which it is attached; the tag is capable of providing this information when requested. So, a tag's functionality is to store the information about the item to which it is affixed, to receive and process the request for this information, and to send a response to this request that carries the information about the item. To support this functionality, a tag has a chip and an antenna housed on a substrate.

To meet a wide spectrum of application requirements, tags come in different kinds. Based on how they communicate, tags are of two types: active tags that can initiate communication and passive tags that cannot initiate communication. Based on the features they offer, the tags are classified into six classes: class 0, class 1, class 2, class 3, class 4, and class 5. Class 0, 1, and 2 tags are passive tags, whereas class 3 tags are semipassive. Class 4 and 5 tags are active tags. Class 0 tags are read-only; class 0+, 1, and 2 tags are write-once, read-only; and class 3, 4, and 5 tags are readable and writable.

Before the tags can be placed on items, they are turned into various kinds of labels, depending on the varied needs of applications. This gives rise to a different way of categorizing tags. Some examples of these forms of tags are smart labels, inserts, RFID-enabled tickets, and tie-on tags. The main goal in placing a tag on an item is that it should be readable by a reader. A reader can only read a tag in a limited area around it, called an *interrogation zone*, discussed in the next chapter.

Exam's Eye View

Comprehend

- ☑ A tag is affixed to an item that needs to be identified and tracked. It contains the item's identification and possibly more information.
- ☑ A tag consists of three main components: a chip, an antenna, and a substrate.
- ☑ A tag's size is limited by the antenna size.
- ☑ A tag's read range increases with the increase in frequency.
- ☑ The defining difference between active and passive tags is that active tags can initiate communication whereas passive tags cannot.
- ☑ When you do not know the tag orientation or you have no control over it, you should use the circularly polarized reader antenna because it can read horizontal tags, vertical tags, and tags aligned to angles between horizontal and vertical.

Look Out

- ☑ Tags and readers using the inductive coupling communication technique use inductive coils instead of standard antennas.
- ☑ Both readers and tags have their own antennas.
- ☑ For a given frequency, tag type, and communication technique, the practical read distance of a tag also depends on other factors, such as the regulated maximum radiated power and antenna size.
- ☑ Passive tags can also operate at UHF and therefore offer a relatively larger read range.
- ☑ UHF passive tags use backscattering.
- ☑ Both active and semipassive tags have their own power sources (such as a battery), but semipassive tags cannot initiate communication.

Memorize

- ☑ Class 0, 1, and 2 tags are passive tags, whereas class 3 tags are semipassive. Class 4 and 5 tags are active tags.
- ☑ Class 0 tags are read-only; class 0+, 1, and 2 tags are write-once, read-only; and class 3, 4, and 5 tags are readable and writable.
- ☑ Class 4 and 5 tags are both active tags, and class 5 tags can act as readers.
- ☑ The read range of a tag depends on a number of factors:
- ☑ Tag type: active or passive
- ☑ Operating frequency
- ☑ Maximum power allowed to be emitted by the antenna
- ☑ Communication technique used: inductive coupling or backscattering.

Key Terms

Active tag A tag that has its own power source such as a battery and that can initiate communication by sending its own signal.

Beacon A tag that emits a signal at predetermined intervals. Beacons are mostly used in real-time locating systems (RTLS).

Electronic product code (EPC) A group of coding schemes for tags defined by the standard called Generation 2.

EPC number A number assigned by a manufacturer to identify the specific object to which a tag is attached. This is also called an *object ID (OID)*.

Inlay The combination of antenna, chip, and substrate.

Insert An inlay inserted between a label in the front and an adhesive layer in the back. The adhesive in the back can be permanently mounted, for example, on the inner wall of a tire.

ISM (industrial, scientific, and medical) A group of frequencies, originally reserved for noncommercial uses in industrial, scientific, and medical fields, now generally used for RFID systems.

Kill function Used to disable a tag permanently.

Label A tag attached to an item by an adhesive with the purpose of identifying the item.

Passive tag A tag that does not have its own power source such as a battery and therefore cannot initiate communication.

Read range The maximum distance from which a tag can be read.

Semipassive tag A tag that has its own power source such as a battery but does not initiate communication.

Substrate A support structure (layer) that houses a tag's antenna and chip.

Tag An RFID component attached to an item that needs to be tracked. It contains the information about the item and provides that information on request.

Tag identifier (TID) A code assigned by a manufacturer to uniquely identify a product.

Self Test

A Quick Answer Key follows the Self Test questions. For complete answers and explanations to the Self Test questions in this chapter as well as the other chapters in this book, see **Appendix A**.

1. Which of the following tags typically use inductive coupling for communication? (Choose two.)

   A. LF

   B. HF

   C. UHF

   D. Microwave frequency

2. Which of the following tags do not have a battery? (Choose two.)

   A. UHF passive tag

   B. HF passive tag

   C. LF semipassive tag

   D. HF active tag

   E. HF semipassive tag

   F. LF active tag

3. Which of the following materials are most disruptive for communication by a UFH passive tag?

   A. Plastic

   B. Water

   C. Cardboard

   D. Paper

   E. Metal

4. When a reader cannot read the tag of an item because it is hidden behind another item, this effect is called:

   A. Ghost read

   B. Shadowing

   C. Low efficiency

   D. Reflection

5. Match the items in the second column of the table to the items in the first column.

Tag Type	Tag Class
A. Active	1. Class 0
B. Passive	2. Class 1
C. Semipassive	3. Class 2
	4. Class 3
	5. Class 4

6. Which of the following best describes an RFID insert?
   A. Tag
   B. Label
   C. Smart label
   D. Inlay
7. Low-frequency passive tags have a read range of about:
   A. 50 centimeters
   B. 3 meters
   C. 9 meters
   D. Greater than 10 meters
8. A tag operating at which frequency can be best read through an animal's body?
   A. 915 MHz
   B. 2.44 GHz
   C. 134 KHz
   D. 13.56 MHz
9. Which of the following tag types may have the largest data capacity?
   A. Active
   B. Passive
   C. Semipassive
   D. Class 2

Self Test Quick Answer Key

For complete questions, answers, and explanations to the Self Test questions in this chapter as well as the other chapters in this book, see the Self Test Appendix.

1. C
2. D
3. B
4. D
5. C
6. A and B
7. B
8. D
9. B

Chapter 4

RFID+

Working with Interrogation Zones

Exam Objectives	What It Really Means
1.1 Describe interrogator functionality	Understand the functionalities, capabilities, and features offered by interrogators. You must understand that the core functionality of an interrogator is to collect data from the tags and send it to a host system. Understand the need and use of different types of interrogators such as handheld, fixed-mount, and vehicle-mount interrogators. Also understand that an interrogator can be connected to a host computer via a serial connection or a network connection. You must understand other capabilities as well that the interrogators can offer such as I/O, firmware upgrade, and GUI.
1.2 Describe configuration of interrogation zones	Understand that configuring interrogation zones involve setting up interrogators, and configuring commands and settings. Also understand how to optimize the interrogation zone while configuring it. You must understand the role of system fine-tuning, tag travel speed, and monostatic and biststic antenna configurations in optimizing the interrogation zone.
1.4 Given a scenario, solve dense interrogator environment issues (domestic/international)	
1.3 Define anti-collision protocols (e.g., number of tags in the field/ response time)	Understand how multiple tags and multiple interrogators can create dense environments. You must understand both kinds of dense environments—tag dense environment and interrogator dense environment— and the corresponding collisions they cause: tag collisions and reader collisions. You must know how Aloha-based protocols and tree-based protocols solve the collision problems and how they are different from each other.

Introduction

An RFID system is based on communication between an interrogator and a tag. The tag is attached to an item that needs to be identified and tracked, and it contains the information about the item such as its identification. The interrogator's job is to collect that information from the tag and send the information to a host computer, where it could be used. For an interrogator to be able to communicate with a tag, the tag must be within a certain area around the interrogator, called the *interrogation zone*. Multiple interrogators and tags can create a crowded environment called a *dense environment* in which things (interrogator zones and signals) can run into each other. Therefore, you need to configure the interrogation zone in an optimal way.

So, the central issue in this chapter is the interrogation zone. To be able to put your arms around this issue, you will explore three avenues: functionality of an interrogator, dense environments, and configuring and optimizing interrogation zones.

Understanding an Interrogator

An *interrogator* is the RFID component that collects information from tags and sends it to a host system. The process of collecting the information from the tags is called *reading the tags*, and for this reason an interrogator is also called a *reader*.

As you know from Chapter 1, the goal of an RFID system is to identify and track items, which is accomplished by tagging the items with tags and collecting the information about the items from the tags. As Figure 4.1 depicts, an interrogator is at the center of this action. From the perspective of an interrogator, the information collection process is performed as follows:

1. The interrogator gets a request for information from the host system.
2. The interrogator sends the request for information to a tag within its interrogation zone.
3. The tag responds with the requested information.
4. The interrogator sends the collected information to the host system.

Figure 4.1 The Role of Interrogator in the Information Collection Process

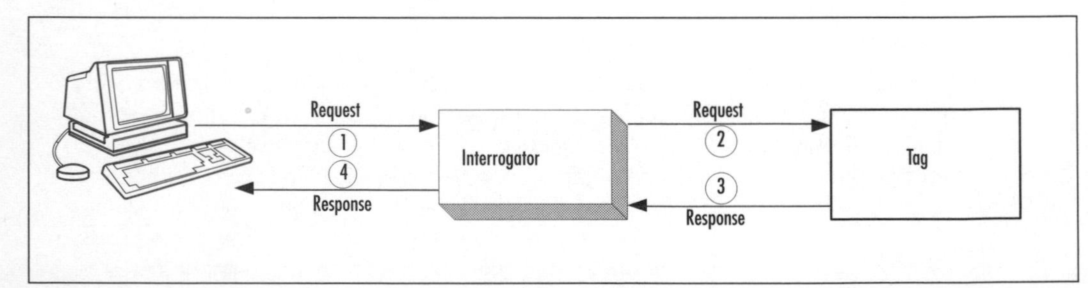

Now that you can appreciate the significance of the role that an interrogator plays in an RFID system, you can ask the following three questions to learn more about it:

1. What is an interrogator made of?
2. What is the functionality of an interrogator?
3. How does an interrogator communicate?

What an Interrogator Is Made Of

An interrogator is composed of the following components:

- A RF module, also called a *transceiver*, which modulates outgoing RF signals and demodulates incoming RF signals
- A signal processing and control unit
- A coupling element that communicates with the tags via RF signals; this is essentially an antenna
- An interface to communicate with the host system: to receive a request for information and to send back the requested information

With these basic components, interrogators come in various types.

Interrogator Types

Interrogators come in various types to meet varied application requirements. All these types are categorized into the following two classes:

- **Read-only** Reading information stored (programmed) in the tag is what an interrogator is basically made for. All those interrogators that can only read the information from the tag and cannot write the information to the tag are called *read-only interrogators*.
- **Read and write** Interrogators that can write information into a tag in addition to reading the information from the tag are called *read and write interrogators*.

CAUTION

Not all interrogators have the write capability. For information to be written to a tag, the interrogator should have the capability to write and the tag should allow the writing.

Both read-only and read/write interrogators come in various types, described in the following sections.

Fixed-Mount Interrogators

Fixed-mount interrogators are fixed-position interrogators mounted at specific locations though which the tagged items are expected to pass. Conveyors, dock doors, and retail store checkout points are some examples of such locations. Any tagged item that passes through the interrogation zone (the area around the interrogator) is scanned—that is, the interrogator reads the information from the tag attached to the item. The advantage of a fixed-mount interrogator is that the tags are read (in other words, the items are scanned) automatically. The disadvantage of a fixed-mounted interrogator is the possibly harsh environment that comes with the location where the interrogator is mounted. Because the reading is automated and because the interrogator is fixed, environmental conditions such as temperature, moisture, vibrations, and materials such as metals can pose challenges. You must take note of these conditions while mounting an interrogator—for example, position an antenna away from metals.

NOTE

Water and material with water content can absorb energy from RF signals and thereby weaken them, and metals can reflect RF signals and thereby change their directions. Both of these factors (weakening the signal and changing the signal direction) decrease an interrogator's efficiency in reading tags.

You might think that you can largely avoid harsh environmental conditions and gain more flexibility if you could get a mobile interrogator. Such devices are available; let's take a look.

Handheld Interrogators

Handheld interrogators are mobile (portable) interrogators, and therefore they contain all the basic elements, including antenna and application software, in one device. The information collected from the tags is stored in the interrogator and later transferred to a data processing system, if the application requires it.

Handheld interrogators offer maximum flexibility. A user can bring the interrogator close to the tagged item and collect the information. So, these interrogators are designed with near-field read/write capabilities. In other words, a handheld interrogator's read range (the maximum distance from which an interrogator can read a tag) is less than that of a mounted interrogator. Handheld interrogators can be used for applications such as tracking and scanning items in medical, office, and retail environments.

NOTE

A handheld interrogator typically supports only one antenna.

Now you get more ambitious and ask the question: Why can't I get the best of both worlds—a mounted interrogator that can be moved? Well, they are available too: they're called vehicle-mount interrogators.

Vehicle-Mount Interrogators

Vehicle-mount interrogators are mobile mount interrogators that can be mounted on a vehicle such as a forklift. Because it's mobile, you can cover a lot more area with a vehicle-mount interrogator than a fixed-mount interrogator. In addition, its read range is larger than that of a handheld interrogator. Because it's inside a vehicle, you can easily add a printer and other peripherals to the system and communicate with the host system wirelessly, for example, using wireless networks based on the 802.11 protocol. A disadvantage of the vehicle-mount interrogator is that it might have to work in the vicinity of metallic materials. This could pose a challenge because metals can reflect the RF signal.

As you can see, the components and types of interrogator are there to serve its functionality, which we explore next.

What an Interrogator Is Good For

Interrogators come with various functions and capabilities that can vary depending on the application for which the interrogators are designed. The functions and capabilities interrogators offer fall into three main categories: communication with the host computer, communication with tags, and operational capabilities.

Communication with the Host Computer

Depending on the application, the data the interrogator collects from the tags may be used by some application such as inventory control. In such cases, the interrogator needs to send the collected data to a host computer to which it is connected. The connection may be a serial connection or a network connection. In a serial connection, the interrogator is connected to the host computer just like a peripheral is locally connected to a computer. In a network connection, the reader is connected to a network to which the host computer is also connected. The network connection may be through a cable or it could be wireless, as in the case of a handheld reader.

Communication with the Tags

Communication with tags involves the following functionalities:

- **Encoding and decoding information** The interrogator communicates with the tag wirelessly by sending and receiving RF waves. It codes the data (information) into an RF carrier signal and transmits this signal into the free space. This signal is to be received by a tag in the interrogator zone. It also receives the response signal from the tags and decodes the information from it. The frequency used for this communication between the interrogator and the tag, called *operating frequency*, varies depending on the applications, standards, and regulations.
- **Powering the passive tags** Tags that do not have their own power source (battery) are called *passive tags*. They get power from the signal they receive from the interrogator and use that power for their operation (i.e., to power up their circuitry) and for composing the response signal that they send back.
- **Reading and writing the tags** Reading the tag—that is, getting the information from the tag about the tagged item—is the minimum functionality of an interrogator. Some interrogators also have the capability of writing information to the tag. However, this can happen only if the tag allows it. That is, the tag must be a read/write tag—that is, a writable tag. Writable tags can allow an interrogator to write new data, modify existing data, or delete the data altogether.

Operational Capabilities

Operational capabilities enable the interrogator to integrate into the RFID system and offer some features. Interrogators can offer three main operational capabilities:

- **Firmware upgrade** Firmware is a software program embedded in a device that configures its basic functionality when the device is powered up. It consists of software instructions stored in nonvolatile memory—the memory that survives even if the device is powered off. This memory is basically a chip called a *read-only memory (ROM) chip*. In the earlier days of personal computing, if you needed to change the instructions on a ROM chip, you would need a new chip because you could not write to it, hence the name *read-only*. However, these days most firmware chips are upgradeable—that is, you can change the instructions on the chip. They can still be called ROM chips for historical reasons. However, the memory on such chips is still nonvolatile, but the content can be upgraded. If the interrogator has upgradeable firmware, it can be upgraded to, for example, new functionalities and new standards and protocols. This capability is important, given that RFID technology and standards are continually evolving.
- **Graphical user interface (GUI)** Graphical user interfaces offer a convenient way to interact with a device—for example, to tell it to do something

and then receive the results (output) of its work. For instance, an open window on your computer is part of the GUI that allows you to interact with the computer. Various interrogators offer various GUI options, depending on the applications. For example, some interrogators offer an HTTP GUI, which means that you can interact with the interrogator through the World Wide Web using your Web browser; others offer only a local GUI. Either way, you can configure and manage the interrogators using the GUIs they offer.

- **I/O capability** You might need to control (or use) the interrogator from another device, or you might need your interrogator to control (or use) another device. In both cases you will need to connect the other device to the interrogator through a port. These ports are called *input/output (I/O) ports* and the devices are *I/O devices*. An example of an input device is a device called an *electronic eye*, which turns on the reader when it senses that an object has entered the interrogation zone. An example of an output device is a *light stack* that signals when a tag has been read.

NOTE

The underlying functionality of I/O capability is that an event can enable the interrogator to do something (input), and the interrogator can create an event in response to the information received from a tag (output).

You will learn more about the communication between an interrogator and a tag throughout the book. An interrogator also communicates with the host computer.

Communicating with the Host

In an RFID system, an interrogator collects information from tags and sends it to the host computer, where it can be used by an application such as inventory system. To be able to send the information to the host computer, the interrogator must be connected to it. The connection could be a serial connection through a serial port or a network connection through a network card (interface) such as Ethernet.

Serial Connections

A serial *connection* consists of a serial port on the reader, a serial port on the host computer, and a cable directly connecting the two serial ports. The data travels though the cable as a stream of bits, one bit at a time, sequentially. The standard protocol used for serial communication in most readers is RS-232, the same protocol that is typically used to send data from your keyboard (as you type) to your computer. Serial connections have the following advantages:

- Low cost
- A reliable and locally managed communication link

However, serial connections also have the following disadvantages:

- The flexibility about the location of the host computer relative to the reader is limited by the cable length.
- Depending on the locations of the readers and the serial ports available on the host computers, you will need multiple host computers for multiple readers.
- If the readers in your RFID system have no network connections, you will need to be physically there to manage them.

Depending on the size of your RFID system, the serial connections can result in higher cost and significant system downtime. The solution to this problem is replacing the serial connections with network connections.

Network Connections

A *network connection* is made through a network card, also called an *Ethernet card* or *interface*. The reader and the host computer are connected to the network through network interfaces such as Ethernet cards and use the TCP/IP protocols to transfer data. For this reason this connection is also called a *TCP/IP connection.*

NOTE

Transmission Control Protocol/Internet Protocol (TCP/IP) is a suite of protocols used by all computers connected to the Internet to communicate with each other. You can think of the Internet as a big network of computers and other devices connected to it.

Some of the protocols included in the TCP/IP protocol suite are described in Table 4.1.

Table 4.1 Some Protocols in the TCP/IP Protocol Suite

Protocol	Description
DHCP	Dynamic Host Configuration Protocol; used on a network to automatically provide an IP address to a computer when it is booted.
ICMP	Internet Control Message Protocol; used by the routers on the Internet to report errors in communication.

Continued

Table 4.1 continued Some Protocols in the TCP/IP Protocol Suite

Protocol	Description
IP	Internet Protocol; used to define IP addresses for devices and to send data to other devices and receive data from other devices.
TCP	Transmission Control Protocol; used for reliable communication with a specific application on a destination device. For example, the recipient will send the acknowledgments to the senders on receiving the data, and if the data does not reach the destination, it will automatically be retransmitted.
UDP	User Datagram Protocol; used for simple but unreliable communication. No acknowledgments and retransmissions are supported.

An interrogator connected to the network using TCP/IP must have a network address called an *IP address*. Network connectivity offers the following advantages to your RFID system:

- There's no need for a cable between an interrogator and a host computer.
- The interrogators can be connected to the network through network cables or wirelessly.
- The system requires a smaller number of host computers.
- You can manage the RFID system remotely.

For a good-sized RFID system, the advantages of a networked system outweigh its disadvantages. However, you should be aware of the possible disadvantages of a networked RFID system:

- Your system becomes vulnerable to all the security risks that a network poses. Of course, security solutions are available.
- A network shutdown will bring the whole system down.
- You need network administrative skills to run a network.

A network of interrogators is an RFID system with multiple interrogators. Each interrogator in your RFID system offers an interrogation zone, and it will attempt to read all the tags passing through (or sitting in) the interrogation zone. For an RFID system with multiple interrogators, there will be multiple interrogator zones, with each interrogator reading multiple tags. This situation can offer what is called a *dense environment*.

Dealing with Dense Environments

Interrogators and tags are two main components of an RFID system. When an RFID system contains multiple tags and interrogators, a condition called a *dense environment* can arise. There are two kinds of dense environments:

- **Dense interrogator environment** A *dense interrogator environment* is an area in which multiple interrogators are operating in close proximity to one another.
- **Dense tag environment** A *dense tag environment* is an area in which multiple tags are in the interrogation zone of an interrogator so that more than one tag can get the same signal from the interrogator.

Dense environments can hamper RFID system performance through effects such as collisions.

Understanding Collisions

What can you expect in a dense (crowded) environment? Yes, you are right: collisions. Corresponding to the two kinds of dense environments are two kinds of collision: reader collisions and tag collisions.

Reader Collisions

Reader collisions occur in a dense interrogator environment. In this environment, the interrogation zone (coverage area) of one interrogator overlaps with the interrogation zone of another interrogator. This overlap causes the following two problems:

- **Multiple reads** More than one reader whose interrogation zones overlap can read the same tag. Depending on the application, these duplicate reads can cause problems. As an analogy, think of counting something multiple times when it's supposed to be counted only once. One of the solutions to this problem is to program the RFID system so that a tag with a given unique ID is read only once.
- **Signal interference** When the interrogation zones of two readers overlap, the signals from the two readers traveling in the overlap area at the same time can collide with each other. This is called *signal interference*. One of the solutions to this problem is that the readers use the time division multiple access (TDMA) technique, according to which the readers read at fractionally different times, thereby reducing the probability of collisions.

Tag Collisions

A *tag collision* occurs when two or more tags try to respond to an interrogator's request for information at the same time. Why would they do that? Because they all were in the interrogation zone, so they all received the request the interrogator sent. The multiple responses will confuse the interrogator and could make it unable to identify any of the responding tags and thereby the tagged items.

NOTE

The dense tag environment also creates a *shadowing effect*, which is a situation in which a tagged item blocks the reader signal from reaching another tagged item hiding behind it. Therefore, the hiding item can never be read and is said to be *shadowed* by the item in front of it.

So, the dense environments create collision problems, which can be addressed by so-called anticollision protocols.

Anticollision Protocols

Where there is a problem, there is (or should be) a solution. A solution to the collision problem is offered by the anticollision protocols, which fall into two categories: aloha-based protocols and tree-based protocols.

Aloha-Based Protocols

The basic goal here is to read one tag at a time. *Aloha-based protocols* accomplish that by using the following two schemes:

- **Time-slotted aloha** In the *time-slotted aloha* scheme, an interrogator keeps periodically sending a request for an ID. Such an interrogator is called a *beacon*. When a tag receives the request, it randomly selects a time slot in which it responds with its ID. If the interrogator recognizes the ID, it starts communicating with that tag to get the required information. After the interrogator is done communicating with this tag, it again starts sending out the request commands that another tag can respond to, and so on. If two or more tags get the same request command from the interrogator, the hope is that the random selection algorithm will generate different time slots for their responses, thereby avoiding the collision. Note that it's possible that the two tags can select (randomly) the same time slot. In this case there will be a collision. So, this approach reduces collisions but does not eliminate them.

- **Frame-slotted aloha** The *frame-slotted aloha* scheme is an extension of the time-slotted scheme. Instead of randomly selecting a time slot, a frame of multiple time frames is configured, and a given tag can only respond in a specific time slot within the frame. This further reduces the probability of collision.

There are two problems with the aloha-based protocols:

- They cannot completely eliminate collisions.
- In the aloha-based protocols, a tag might not be identified for a long time because other tags keep selecting time slots earlier than that of this tag. This situation is called *tag starvation*.

Tree-based anticollision protocols offer a solution to the tag starvation problem.

Tree-Based Protocols

Tree-based protocols use the algorithm that splits the group of colliding tags into two subgroups iteratively until the reader recognizes the tag IDs without collisions. This can be done in two different ways, which gives rise to two tree-based protocols:

- **Binary tree protocol** To support the *binary tree protocol,* the tags are required to manage a counter and have a random number generator. The colliding tags are split according to a number that they randomly select. The tags that select 0 transmit their IDs to the interrogator. If multiple tags select 0 and hence respond, the interrogator keeps walking down the tree until only one tag responds. When that happens, the interrogator establishes communication with that tag to get the required information.
- **Query tree protocol** The *query tree protocol* uses the algorithm, following which the interrogator sends a query with a prefix and the tags that have the ID to match the prefix respond.

Tree-based protocols solve the tag starvation problem, but they can create long identification delays. So, the underlying goal of all anticollision protocols is to select only one tag at a time that the reader can communicate with. However, for a tag to be read by an interrogator, it must be in the interrogation zone. The interrogation zones need to be set up and configured.

Configuring Interrogation Zones

The *interrogator zone* is the area around an interrogator within which it can successfully communicate with a tag. In other words, when a tag enters an interrogation zone, it can be interrogated by the interrogator. From the perspective of a passive tag, the interrogator zone is the area in which an interrogator can provide enough energy to power up the passive tag and receive information. Passive tags outside the interrogation zone do not receive enough energy from the interrogator to reflect a signal.

CAUTION

The interrogation zone is sometimes also called the *read field* or the *reader field*.

Interrogation zones need to be configured, which involves setting up readers at specific locations where the large number of items pass through. These points are called *choke points*.

The definition of a successful configuration of an interrogation zone will be influenced and partly determined by the following two factors:

- Business process flow
- Site assessment, including physical infrastructure, discussed in Chapter 7

However, the following are the common factors that you should consider to successfully configure the interrogation zone:

- The read rates required by the tag traffic
- Power required by the interrogator and the tags
- Distance required or available between the interrogator and the tags

Configuring an interrogation zone involves the following:

- Configuring interrogator commands
- Configuring interrogator settings
- Adjusting the read power of the interrogator to an optimal value

Configuring Interrogator Commands

Interrogator commands are usually issued on the host computer, either by an application or using a GUI. Some of the common interrogator commands are described in Table 4.2.

Table 4.2 Some Common Interrogator Commands and Their Usage

Command	Action
KILL	Disable the tag permanently
LOCK	Disable writing to the tag
QUERY	Initiate communication with the tag
READ	Get information from the tag
WRITE	Write the ID or other information to the tag
UNLOCK	Enable writing to the tag—that is, remove the write protection on the tag

As you can see in Table 4.2, some commands such as *LOCK* and *UNLOCK* configure certain behavior or capability in the interrogator.

CAUTION

One use of the *KILL* command is to address privacy concerns. However, remember the other side of the coin: It can also be used maliciously to disrupt the system.

The *KILL* command, once issued, prevents a tag from communicating back to the reader, and it appears to the reader as inoperative. Why will you use the *KILL* command? One reason is to manage the tag population. Other commands that you can use to manage the tag populations are:

- ***SELECT*** You can use this command to determine which groups of tags will respond to this interrogator. For selection purposes, you can group the tags by characteristics such as manufacturer code. By isolating only certain groups of tags that the interrogator should care about, you increase system performance, because now that interrogator has fewer numbers of tags to deal with and to sort through. Once you have grouped together some tags that the interrogator should care about, you can identify the individual tags in the group.
- ***INVENTORY*** This command is used to identify an individual tag in a group. After an individual tag has been identified, the interrogator can access it.
- ***ACCESS*** This command is used to access the individual tag in a group. Once the interrogator has access to an individual tag, it can deal with it, such as reading the information from it, writing information to it, killing the tag, and so on.

The tag population is controlled to optimize the interrogation zone and improve system performance. In addition to these commands, there are some other settings that you can configure.

Configuring Interrogator Settings

Different interrogators offer different settings and features that you can configure. Some of these settings are described in Table 4.3.

Table 4.3 Some Settings That You Can Configure for Readers

Setting	Description
Event notification	When enabled, this setting provides notification when a certain event occurs, such as a tag entering the interrogation zone or the number of tags in the interrogation zone exceeding a threshold value.
Filtering	Sets specific filters and associates them with specific read points. This allows you to tell the interrogator to be interested only in certain kinds of tags and to ignore other kinds.
Host management	Authorizes the listed host computers, called *trusted hosts*, to communicate with the interrogator.
Reader communication	Allows you to set communication-related settings.
Reader operation	Allows you to rename, enable, or disable the interrogator.
Read point zone	Allows you to logically group two or more read points for management purposes.
User management	Allows you to add users who can then be given access and management rights.

Some available software applications will help you configure, monitor, and manage the readers in your RFID system. Following are some additional features offered by such applications:

- **Reader status** You can get the status of a specific interrogator by issuing the specific reader status command offered by the application that will typically display the following information:
    - Information about the interrogator's kernel
    - Information about the read points and antennas attached to the interrogator
    - How long the reader has been up and running
- **Overall status** You can also get details on the overall status of the RFID system, such as the following:

    - Total number of readers connected to the system
    - Readers enabled for reading
- **Scan control** You can use this feature to scan the read points and enable and disable them.

The other interrogator zone-related configuration tasks that you are allowed to perform are the following:

- Select RFID protocols to be used by the interrogator, such as anti-collision protocols
- Modify the configuration options for the protocol that you selected
- Set the RF mode
- Adjust output power
- Enable tag alerts

You can also configure interrogator commands and settings to optimize the interrogation zone.

Optimizing Interrogation Zones

You want your RFID system to be reliable, robust, and performing at its peak. The reliability, robustness, and performance are built into the system components, but you can optimize them by fine-tuning the way the components work together. One way you can make your system more reliable and robust and yield peak performance is by optimizing the interrogation zones. Optimizing an interrogation zone involves correctly setting up the system for the given environment and application in which the interrogator will function. Some of the factors that you need to consider for optimizing the interrogation zones are discussed in this section.

The Network Factor

A network has a limited bandwidth for communication—that is, for transferring data. All the devices on the network share that bandwidth. An RFID reader can typically read hundreds of tags per second; each read cycle for each reader consumes bandwidth. Uncontrolled readers can slow the network by consuming large shares of bandwidth. Therefore, network traffic must be monitored and managed. In addition, to optimize the system you are setting up, you should carefully consider the bandwidth and the number of readers that will share that bandwidth on your network.

NOTE

Bandwidth is the width of a band of electromagnetic frequencies used for transferring data. It is a measure of how fast data can be transferred on a given transmission path and determines the total data transmission rate that the path can offer. The basic units of bandwidth are Hertz, abbreviated Hz (cycles/sec), in the analogous world and bits/sec in the digital world.

Operation Mode

Your interrogator's performance can depend on the mode in which the interrogator is communicating. An interrogator can communicate in one of the following two communication (or *operational*) modes:

- **Half duplex** This is the mode in which data transfer between two devices can occur in only one direction at a time. That means that a reader operating in half-duplex mode can either send signals to tags or receive signals from a tag but cannot send and receive at the same time. An interrogator operating in half-duplex mode is configured for a *monostatic antenna*, which means that the interrogator uses the same (one) antenna for sending and receiving signals. A monostatic antenna is also called a *patch antenna*. A monostatic antenna configuration offers a smaller coverage area.
- **Full duplex** This is the mode in which data can be transferred between two devices in both directions simultaneously. That means that a reader operating in full-duplex mode can send a signal to the tag and receive a signal from a tag at the same time. An antenna configuration for the reader that offers this functionality is called a *bistatic antenna*, which means that the interrogator will use one antenna for sending signals and another antenna for receiving signals. A bistatic antenna configuration offers a wider coverage area.

Reader-to-Reader Interference

As explained earlier in this chapter, if multiple readers are too close to each other, their interrogation zones can overlap. This will cause collisions and interference between the signals from different readers. The anticollision protocols as solutions to collisions have already been discussed in this chapter. You can also consider the following solutions to avoid interference problems:

- Position the interfering antennas away from each other.
- Reduce the interrogator power.

- Set the interfering interrogators to operate on different frequencies.
- Program the RFID system so that a tag with a given unique ID is read only once. This will solve the multiple reads problem.
- Set the readers to use the TDMA technique, according to which the readers read at fractionally different times, thereby reducing the probability of collisions.

You can also properly tune your reader to improve system performance.

System Performance and Tuning

You can tune and configure certain characteristics of a reader, such as power output and protocols, to optimize its performance in a given environment and application:

- **Power output** You can fine-tune an interrogator by adjusting its output power. Remember that always using the maximum allowed power output may not be the best choice for your system. So you might need to optimize the power settings for a given environment and application. You should evaluate the physical quantities such as attenuation, ERP, and free space loss to adjust the power settings for an interrogation zone.
- **Protocol configuration** Different tags may support different communication protocols. To read all kinds of tags, a reader typically has to support and execute several protocols because the reader does not know ahead of time what protocol the incoming tags might be supporting. This takes time and slows down the reader. However, when all the tags are coming from the same location and support a specific protocol, you can configure the reader for that specific protocol only. This will improve the reader's performance. However, keep in mind that you cannot do this for a location where the tagged items come from different places and the tags might be supporting different protocols.
- **Read cycle rate** A *read cycle* is a scan for RFID tags performed by a reader. The reader can run read cycles periodically or on demand. After a read cycle, the reader returns (for example, to the host computer) a set of observations—for example, a set of IDs for the tags that were read. For peak performance, the read cycle rates must be optimized. For instance, if you know that only a certain kind of tag will enter a given interrogation zone, you configure the interrogator to scan (look) for only that kind of tag. This will reduce the overall scan time and thereby improve performance.

The Tag Travel Speed

The term *tag travel speed* refers to the speed of the tagged item or the speed of the platform such as a conveyor on which the tagged items are placed. The travel speed determines the following:

- The duration for which a tag will stay in the interrogation zone
- The number of tags that will pass through the interrogation zone in a given duration and will need to be read in that duration

The higher the tag travel speed, the higher the reader read speed that is required for an optimal interrogation zone. So, to determine the optimal read speed for the reader, you need to know how many tags will pass through the interrogation zone per unit of time. If you decrease the tag speed too much, the tag read rate will decrease as well, and if you increase the tag speed too much, the reader will miss reading some of the tags, which will hamper accuracy. Therefore, to optimize the interrogation zone, you need to strike a balance between tag speed and the reader's read speed.

The three most important takeaways from this chapter are the following:

- The job of an interrogator is to collect information from the tags in its interrogation zone and send that information to a host computer, where it can be used.
- Multiple tags in an interrogation zone create a dense tag environment, which causes tag collision—that is, multiple tags try to respond to an interrogator at the same time. Overlapping interrogation zones create a dense interrogator environment, which causes reader collision—that is, multiple readers try to read the same tag and their signals interfere.
- You should configure a reader to optimize its interrogation zone. The strategy for optimization may depend on the environment and the application.

Summary

The core functionality of an interrogator (also called a *reader*) is to collect information from tags and send that information to a host system. For a tag to be read successfully, it must be in the area around the interrogator, called the *interrogation zone*. Multiple tags in an interrogation zone can create an environment called a *dense tag environment*, which can cause tag collisions—that is, multiple tags try to respond to an interrogator at the same time. Multiple interrogators close to each other may create overlapping interrogation zones, called a *dense interrogator environment*. The dense interrogator environment causes reader collisions, which means multiple readers try to read the same tag and the signals from multiple readers interfere with each other. The collision problem is addressed by anticollision protocols, which fall into two categories: aloha-based protocols and tree-based protocols. Aloha-based protocols create the tag starvation problem—that is, a tag may have to wait for very log time before it could be identified; whereas tree-based protocols solve the starvation problem, but they can create long identification delays in general.

Interrogation zones need to be configured, a process that includes setting up readers and configuring reader commands and options. While configuring an interrogation zone, you must try to optimize it for performance and application requirements. The definition of successful configuration and optimization partly depends on the environment in which the RFID system works and the application. In this chapter, we talked about collision protocols. Protocols and standards are important for the smooth advancement of any industry. We discuss RFID standards in the next chapter.

Exam's Eye View

Comprehend

- ☑ An interrogator collects the information about an item from the tag attached to that item and sends the information to a host computer, where it can be used.
- ☑ Passive tags use the power from the signal the interrogator sends to power up their circuitry and to compose the response signal.
- ☑ Fixed-mount interrogators are set to scan automatically and are built to withstand harsh environmental conditions.
- ☑ Handheld interrogators offer greater flexibility because they can be moved to any place and are designed to have shorter read range.
- ☑ A dense interrogator environment creates reader collisions due to overlapping interrogation zones, whereas a dense tag environment creates tag collisions.

- ☑ The greater the tag speed in the interrogation zone, the greater the read speed required from the reader.

Look Out

- ☑ Of course, all interrogators can read tags, but some interrogators can also write to tags.
- ☑ Not all interrogators have the write capability. For information to be written to a tag, the interrogator should have the capability to write, and the tag should allow the writing.
- ☑ A dense tag environment creates a shadowing effect—that is, a tag blocks another tag from being read.
- ☑ The ***KILL*** command on an interrogator is used to disable a tag and not to disable the interrogator.
- ☑ If you increase tag speed too much, the reader will miss reading some tags, whereas if you decrease tag speed too much, the read rate will be low.

Memorize

- ☑ The main challenge of a vehicle-mounted interrogator is the metal that can reflect the RF signal, resulting in reduced read range and read rates.
- ☑ Firmware is a software program stored in nonvolatile memory. Some interrogators offer the capability of upgrading the firmware.
- ☑ Portable interrogators use wireless technology to send data to host computers, and fixed-mount interrogators can use serial connections or a network connections. The network connection may be cabled or wireless.
- ☑ The standard protocol used for serial communication in most readers is RS-232.
- ☑ Aloha-based anticollision protocols create the tag starvation problem. Tree-based anticollision protocols solve this problem, but they can create long identification delays.

Key Terms

Antenna The device used to transmit and receive signals such as radio waves. Both a reader and a tag have their own antennas through which they communicate with each other.

Bandwidth The width of a band of electromagnetic frequencies used for transferring data. It is a measure of how fast data can be transferred on a given transmission path and determines the total data transmission rate that the path can offer. The basic units of bandwidth are Hz (cycles/sec) in the analogous world and bits/sec in the digital world.

Bistatic antenna configuration A configuration in which an interrogator will use one antenna for sending signals and another antenna for receiving signals. This configuration enables the full-duplex communication mode.

Choke point A specific location through which lots of items pass. The interrogation zones are usually set up at choke points.

Data transmission rate Actual rate, in bits/sec, at which data is being transmitted over a communication line from one device to another. The transmission can be wireless as well.

Dense interrogator environment An area in which multiple interrogators are operating in close proximity to one another.

Dense tag environment An area in which multiple tags are in the interrogation zone of an interrogator so that more than one tag can get the same signal from the interrogator.

DHCP Dynamic Host Configuration Protocol, used to automatically (dynamically) provide IP addresses to devices connected to a network.

Firmware A software program embedded in a device that configures its basic functionality when the device is powered up.

Full duplex This is the mode in which data can be transferred between two devices in both directions simultaneously.

Half duplex This is the mode in which data transfer between two devices can occur in only one direction at a time.

HTTP Hypertext Transfer Protocol; a protocol on which the World Wide Web is based. Web browsers and Web servers use this protocol to interact with each other.

ICMP Internet Control Message Protocol; used by routers on the Internet to report errors in communication.

Interrogation zone The area around an interrogator within which it can successfully communicate with a tag. In other words, if a tag enters the interrogation zone, it can be interrogated by the interrogator.

Interrogator The RFID component that collects information from tags and sends it to a host system. The process of collecting the information from the tags is called *reading the tags*, and for this reason an interrogator is also called a *reader*.

IP Internet Protocol; used to define IP addresses for devices and to send data to communicate with a destination device.

Monostatic antenna configuration The configuration in which an interrogator uses the same (one) antenna for sending and receiving signals. This configuration enables only the half-duplex mode of communication.

Operating frequency The frequency of the radio waves that the interrogator and the tag use to communicate with each other.

Read cycle A scan for RFID tags performed by a reader. The reader can run read cycles periodically or on demand.

Reader The RFID component that collects information from tags and sends it to a host system. The process of collecting the information from the tags is called *reading the tags* or *interrogating the tags*. For this reason, a reader is also called an *interrogator*.

Read rate The number of tags that a reader can read per unit of time. Sometimes, read rate is also used for maximum data transfer rate—that is, the maximum rate at which data can be read from a tag, expressed in units of bits/sec.

Network connection A connection made between two devices by connecting them to the same network through their network interfaces (cards). For a network connection, a device must have an IP address.

Reader collision An interference in communication that occurs because two or more interrogation zones are overlapping. This situation is a result of a dense interrogator environment.

Serial communication The process of transferring data from one device to another sequentially, one bit at a time.

Serial connection A connection set up between two devices by connecting their serial ports through a cable.

Tag collision An interference in communication that occurs because two or more tags try to respond to an interrogator at the same time. This situation is a result of a dense tag environment.

Tag starvation A situation created by aloha-based anticollision protocols in which a tag has to wait for long time before it can be identified by a reader.

TCP Transmission Control Protocol; used for reliable communication with a specific application on a destination device.

UDP User Datagram Protocol; used for simple but unreliable communication with applications on other devices.

Self Test

A Quick Answer Key follows the Self Test questions. For complete answers and explanations to the Self Test questions in this chapter as well as the other chapters in this book, see Appendix A.

1. The *KILL* command on an interrogator is used to:

   A. Shut down an RFID system.

   B. Disable a tag.

   C. Disable the interrogator.

   D. Shut down the host computer.

2. All of the following are functions of an interrogator except:

   A. Communicating with another interrogator

   B. Communicating with a tag

   C. Communicating with a host computer

   D. Storing information about an item to which the interrogator is attached

3. While talking about connecting a reader to the host computer, your manager mentions RS-232. What kind of connection is he referring to?

   A. Network connection

   B. Serial connection

   C. Wireless connection

   D. Parallel connection

4. Which of the following is true about shadowing?

   A. It is a situation created by a dense tag environment in which a tagged item blocks the reader signal from reaching another tagged item hiding behind it.

   B. It is a situation created by a dense interrogator environment in which two readers try to read the same tag.

   C. It is a situation created by a dense tag environment in which two tags try to respond to the same reader at the same time.

   D. It refers to the situation that occurs when a tag is too close to an antenna to be read properly.

5. Which of the following is not an effect of dense environments?

   A. Reader collision

   B. Tag collision

   C. Shadowing

   D. Locking the tags

6. All of the following are the configuration settings of an interrogator except:

   A. Tag population that needs to be read

   B. Tag read rate of the interrogator

   C. Anticollision protocols

   D. Power emitted by the reader

7. Match the items in the second column of the table to the items in the first column.

RFID Command	Action
A. KILL	1. Disables writing to the tag
B. LOCK	2. Disables the tag permanently
C. QUERY	3. Gets information from the tag
D. READ	4. Initiates communication with the tag

8. You know that an interrogator zone at your site always gets items with class 0 tags attached to them. You have noticed that the read rate of this zone is not that great. What can you do to increase the read rate?

A. Increase the power of the interrogator to maximum.

B. Set the interrogator to search for all tag classes except class 0.

C. Set the interrogator to search for class 0 tags only.

D. Set the interrogator to search for all tag classes.

9. You have set up 10 interrogators in a small warehouse. You have observed that some of the data the interrogators are collecting is erroneous. What do you think the most likely problem is?

A. Reader collision

B. Shadowing

C. Low power emission by the readers

D. Operating frequency too high

10. All of the following methods will help prevent overlaps of interrogation zones except:

A. Positioning the interfering antennas away from each other

B. Readers are configured to use different protocols

C. Reducing the interrogator power

D. Setting the interfering interrogators to operate at different frequencies

Self Test Quick Answer Key

For complete answers and explanations to the Self Test questions in this chapter as well as the other chapters in this book, see **Appendix A**.

1. B
2. D
3. B
4. A
5. D
6. B
7. A, B, C, D
8. C
9. A
10. B

Chapter 5

RFID+

Working with Regulations and Standards

Exam Objectives	What They Really Mean
3.3 Recognize regulatory requirements globally and by region. **3.4** Recognize safety regulations/issues regarding human exposure.	You should know the different UF frequencies allocated to the RFID devices in different parts of the world. You must know that the world is organized into three regulatory regions. You must also know in which regions the main countries of the world are included. Know the regulations about specific absorption rate (SAR).
3.1 Given a scenario, map user requirements to standards. **3.2** Identify the differences between air interface protocols and tag data formats.	Understand the importance of RFID standards and their impact on the industry and on users. You must know that the RFID standards fall into two main categories: air interface protocols, which deal with the communication part, and tag data formats, which deal with formatting the information stored on tags. You should also know which protocol is developed by which organization: ISO or EPCglobal.

Introduction

All mature (or maturing) industries have their regulations and standards, and the RFID industry is no exception. Regulations help make the devices and systems in the industry secure and safe, and they help the industry advance without disrupting other industries. Standards are necessary to bring some order and interoperability within a specific industry. Without agreed-on standards, all vendors will manufacture or develop products and devices by following their own rules, and there will be a perfect chaos instead of interoperability. Of course, regulations and standards have their impact on an industry's products as well.

So, the main goal of this chapter is to understand the regulations and standards at work in the RFID industry. To accomplish this goal, we will explore three avenues: RFID regulations, RFID standards, and the impact of these regulations and standards.

Understanding Regulations and Standards

What are regulations and standards, and why do we need them? First, note that regulations and standards are not limited to RFID systems. Let's take a general look at these two concepts: regulations and standards.

Regulations

A regulation, in general, is a legal restriction promulgated by a government administrative agency through rule making and is typically supported by a threat of consequences, such as fine for not following the rules. A regulation is mandated by the government or state as an attempt to produce an outcome that might not otherwise occur or to prevent an outcome that might otherwise occur. Regulations rarely work perfectly; they don't always produce the complete desired outcome or completely prevent the undesired outcome, but they do generally modify what would otherwise take place. Examples of regulations include controlling market entries, prices, wages, pollution, employment for certain groups of people in certain industries, and standards of production for certain goods and services—as well as, of course, the regulations involved in manufacturing RFID devices.

Standards

The term *standard* refers to the way something should be done. When multiple vendors are producing the same product in different ways, the products from those different vendors will not interoperate. If all those vendors followed the same standard, their products would be compatible with each other and would be interoperable. So, in some industries the absence of a standard would mean chaos.

In the context of industries and technologies, *standardization* is the process of establishing a technical standard among competing vendors in a market to bring benefits without hurting competition. As an example, all of Europe uses 230-volt, 50 Hz, AC main grids and Global System for Mobile Communications (GSM) cell phones, and they measure length in meters. An example of global standards is the Internet, which is based on standard protocols. Other examples of global standards are the worldwide standards and drafts for the standardization of power cords developed and maintained by the International Organization for Standardization (ISO), the International Electrotechnical Commission (IEC), and the International Telecommunications Union (ITU).

And then there are standards for RFID devices. One parameter that is regulated in the RFID industry is the frequency at which RFID devices can operate.

Regulating Frequency Usage

A tag and a reader use radio waves of a certain frequency, called their *operating frequency*, to communicate with each other. Radio waves are electromagnetic waves that cover part of the electromagnetic spectrum of frequencies, called *radio frequency spectrum*. Because RFID systems generate and radiate the electromagnetic waves that fall along the radio frequency spectrum, they are justifiably classified as radio systems, and they are regulated as such. However, other radio services were in operation before the arrival of RFID systems. Radio, television, mobile radio services (police, security services, and industry), marine and aeronautical radio services, and mobile telephones are a few to count. Therefore, it is important to ensure that these already existing services are not disrupted or impaired by these newcomers: the RFID systems.

For this reason, regulatory bodies allocate different frequency bands (ranges) to a specific group of devices. RFID systems are available in all the radio frequency ranges: LF, HF, UHF, and microwave. Here is the situation with allocating the specific frequencies to the RFID systems in these RF bands:

- **LF** Most countries have allocated 125 KHz or 134 KHz to RFID devices.
- **HF** Most countries have allocated 13.56 MHz to RFID devices.
- **UHF** Different countries have allocated different frequencies.
- **Microwave** Different countries use different frequencies.

As you can see, the frequencies being used by various countries in the LF and HF ranges are very consistent with each other. However, because RFID systems with operating frequencies in the UHF range are relatively new, no global agreement on the operating frequencies for these devices has been reached yet. UHF RFID systems have evolved at different frequencies in different regions of the world. The absence of a single global organization to develop regulations and standards for RFID technology has prompted countries to adopt their own regulations and standards.

The Regulatory Regions

If the regions of the world use different operating frequencies for RFID systems, an RFID device that works in one region will not work in another region. Because the UHF RFID systems are gaining popularity all over the world, it's desirable to have some uniformity in the operating frequencies for these devices. In an attempt to seek some degree of uniformity for UHF frequency usage, the ITU has organized the world into the following three regulatory regions:

- **Region 1** includes Europe and Africa.
- **Region 2** includes North and South America.
- **Region 3** includes Asia and Australia.

Table 5.1 presents some information such as allocated UHF bands and allowed maximum power emissions, as regulated by the main regulatory bodies in these regions.

TIP

It might sound silly, but it's important, at least from the exam viewpoint, to remember which country belongs to which region. For example, the United States belongs to region 2, not region 1.

Table 5.1 The Three Radio Frequency Regulatory Regions of the World

	Region 1	Region 2	Region 3
Areas covered	Africa, Europe, the Middle East, and the former Soviet Union, including Siberia	North America, South America, and Pacific east of the international dateline	Asia, Australia, and the Pacific Rim west of the international dateline
Main regulatory body	In Europe: European Conference of Postal and Telecommunications (CEPT)	In the United States: Federal Communications Commission (FCC)	In Japan: Ministry of Public Management, Home Affairs, Posts and Telecommunications (MPHPT)
Allocated UHF band	865–870 MHz	902–928 MHz	~950 MHz
Maximum power emission	2W (ERP) = 3.28 W (EIRP)	4W (EIRP)	—

Each country in these regions manages its frequency allocations within the guidelines set by the region's main regulatory body. Table 5.2 shows the RF ranges used for the RFID devices in various countries. This table reflects the following facts:

- The HF RFID devices use 13.56 MHz operating frequency all over the world.
- Most countries have allotted 125 MHz or 134 MHz for HF RFID devices.
- There is better global agreement on operating frequency for the RFID devices in the LF, HF, and microwave ranges than in the UHF range.

Table 5.2 RF Bands Used for RFID Devices in Various Countries

Country	LF	HF	UHF	Microwave
United States	125, 134 KHz	13.56 MHz	902–928 MHz	2.40–2.48 GHz 5.72–5.85 GHz
Europe	125, 134 KHz	13.56 MHz	868–870 MHz	2.45 GHz
China	125, 134 KHz	13.56 MHz	N/A	N/A
India	125, 134 KHz	N/A	865–867 MHz	2.40 GHz
Japan	125, 134 KHz	13.56 MHz	950–956 MHz	2.45 GHz
Singapore	125, 134 KHz	13.56 MHz	923–925 MHz	2.45 GHz

The last column in Table 5.2 shows the regulated maximum power that can be emitted by an RFID device. The power factor is especially important for passive tags because they don't have their own power source and use the power from the reader's signal to run their circuitry and to compose the response signal. Consider an isotropic antenna radiating power uniformly in all directions. That means the power will travel in the form of a sphere. The area of the sphere is directly proportional to the square of the radius of the sphere, which implies that the energy per unit area will be inversely proportional to the square of the distance from the antenna. In other words, RF energy radiated by an antenna dissipates very quickly as the RF wave travels through space. For example, every time the distance from an antenna doubles, the power available reduces by a quarter. You can turn it around to say that the read range of a reader is directly proportional to the square root of the power emitted. Exercise 5.1 demonstrates this concept.

EXERCISE 5.1

You are traveling with a radio that operates in the same frequency as an RFID system. You are in a country where the allowed maximum power emission for this RFID device is 0.5 W. To avoid interference with this device, you have to keep your radio 3 meters away from this device. Now you move to another country where the maximum power emitted by this RFID device is regulated to be 4 W. How far away will you have to keep your radio from this device to prevent interference?

Solution: Given:

$P1 = 0.5$ W, $P2 = 4$ W, $d_1 = 3$ meters

Because the read range is directly proportional to the square root of the emitted power, we can express this relationship in the following formula:

$$d_1 = C \times P1^{1/2}$$
$$d_2 = C \times P2^{1/2}$$

where *C* is a constant of proportionality. This implies:

$$d_2 / d_1 = (P2/P1)^{1/2} \Rightarrow d2 = d1 \times (P_2 / P_1)^{1/2} = 3 \times (4/0.5)^{1/2} = 8.5 \text{ m}$$

Regulating frequencies prevent interference between the operations of different RF devices and provide interoperability among RFID systems. And there is another advantage of RF regulations: safety.

Safety Regulations

The human body exposed to RF radiation absorbs the energy (power) from the RF waves. Safety regulations and guidelines for human exposure to RF fields are necessary because if the RF energy absorption exceeds a threshold value, adverse biological effects could occur. Some of these adverse effects are:

- Changes in cell cycle and cell proliferation
- Changes in the blood-brain barrier that protects the brain from external harmful chemicals and toxins
- Alterations in electric brainwaves

The absorption of RF energy is measured in a quantity called *specific absorption rate (SAR)*, which is a measure of the rate of energy absorbed by (or dissipated in) an incremental mass contained in a volume element of dielectric materials such as biological tissues. The SAR is calculated using the following equation:

$$SAR = C \times E^2/d$$

where:

- *C* is conductivity of the body tissue in S/m (Siemens per meter, where Siemen is just a reciprocal of Ohm, the unit for resistance).
- *E* is the electric field strength in the tissue in *V/m*.
- *d* is the density of the body tissue in Kg/m^3.

In the United States, the FCC has adopted limits for safe exposure to RF energy produced by mobile devices and requires that devices such as cell phones sold in the United States have a SAR level at or below 1.6 W per kilogram, taken over a volume of 1 gram of tissues. In Europe, the corresponding limit is 2 W/kg taken over a volume of 10 grams of tissues. The Institute of Electrical and Electronics Engineers (IEEE) has its own guidelines, as shown in Table 5.3.

Table 5.3 SAR Limits Adopted by Various Regulatory Bodies

Standard	SAR Limit
IEEE	0.2 W/Kg (for the entire body)
FCC (U.S.)	1.6 W/Kg (taken over a volume of 1 gram of tissue)
Europe	2 W/kg (taken over a volume of 10 grams of tissue).

CAUTION

When designing and implementing an RFID system, you must check out and follow the safety guidelines set by local, regional, national, and international organizations for human exposure to RF waves.

So, the regulations in the RFID industry ensure that RFID devices are safe and that they do not disrupt the already existing services. But how about the interoperability of the devices from different vendors—that is, the ability of the devices to function together effectively? Well, that's the job of standards.

RFID Standards

Standardization of its products is one of the important issues that any emerging industry has to deal with. Following are the advantages of having industry standards:

- Because all vendors follow the same standard to manufacture devices, technical standards ensure the interoperability of the devices. This benefits the consumer and helps vendors develop healthy competition.
- Because the standards bodies are not serving the interests of just one vendor, standards generally define the most efficient platform on which an industry can operate and advance.
- Standards generally reduce cost and ease implementation.
- Standards develop consumer confidence in the technology.

Several organizations have been involved in developing standards for RFID technology; the ISO and EPCglobal are the prominent two.

ISO Standards

The ISO is an international standards body composed of representatives from national standards bodies. Founded on February 23, 1947, this organization sets worldwide industrial and commercial standards, which are popularly called *ISO standards*.

The ISO has developed RFID standards in the following areas:

- Identification standards regarding the coding of ID or other information on tags.
- Air interface protocols that define the rules of communication between tags and interrogators.
- Data protocols for the middleware of an RFID system.
- Standards for testing, compliance, and safety.

Some of these standards are shown in Table 5.4.

NOTE

The International Electrotechnical Commission (IEC) is an international standards organization in the area of electrical, electronic, and related technologies. Some of its standards are developed jointly with the ISO.

Table 5.4 Some RFID Standards Developed by the ISO

ISO Standard	Description
ISO/IEC 15961	Information exchange in a radio frequency identification (RFID) system (data protocol for application interface) for item management
ISO/IEC 15962	Data encoding rules and logical memory functions for item management
ISO/IEC 15963	Unique identification for RF tags
ISO/IEC 18000-*i* *i* is an integer: 1, 2, 3 ...	Parameters for air interface communications for different operating frequencies
ISO/IEC 18047-*i* *i* is an integer: 1, 2, 3 ...	RFID device tests methods for different operating frequencies
ISO/IEC 19762-3	Automatic identification and data capture (AIDC) techniques: vocabulary
ISO/IEC 24730-1	Real-time locating systems (RTLS): application program interface (API)

The various values of the integer *i* in the table correspond to different operating frequencies. The air interface protocols define the rules for communication between readers and tags. This includes the rules about the following tasks:

- Data encoding, including modulation and demodulation
- Communication commands to make operations on the tag, such as reading, writing, modifying, and locking data, as well s killing the tag
- Anticollision algorithms

The ISO develops standards in several areas, including computer networking. In the world of RFID standards, there is another player specific to RFID: EPCglobal.

EPCglobal Standards

Here is how EPCglobal came into the picture: The Auto-ID Center at Massachusetts Institute of Technology (MIT), working in conjunction with industry leaders and academic institutions around the world, designed a system to bring the benefits of RFID to the global supply chain. This system comprises the Electronic Product Code (EPC), RFID technology, and the supporting software based on EPCglobal standards, and is referred to as the *EPCglobal Network*. The network includes elements such as EPC, the ID system for EPC tags and readers, and Object Name Service (ONS). The EPCglobal network (or any RFID information network like this) provides the following five main services:

- **Assigning unique identification numbers to items to enable them to be identified** EPC numbers allow item-level tracking.
- **Detecting and identifying items** EPC tags and readers make it possible.
- **Collecting and filtering data** EPC middleware provides services that facilitate data exchange between EPC readers and business information systems such as databases. Only the data about events of interest will be stored.
- **Querying and storing data** This service enables different enterprise applications running at different locations to exchange and share data. That means the trading partners can query and exchange data among themselves.
- **Locating information** This is a lookup and discovery service to locate the repositories for the required EPC data.

EPCglobal Inc. is a joint venture between GS1 (formerly known as EAN International) and GS1 US (formerly the Uniform Code Council Inc.). The organization was set up to achieve worldwide adoption and standardization of EPC technology in an ethical and responsible way. In other words, EPCglobal is leading the development of industry-driven standards for EPC to support the use of RFID in today's trading network environments. The EPCglobal Gen 2 (popularly called Gen 2) standard, approved in December 2004, is likely to form the backbone of RFID tag standards moving forward.

NOTE

The Gen 2 standard is designed to work globally and enjoys the support of major manufacturers.

What is EPC, anyway? EPC is a family of coding schemes for Gen 2 tags. It is designed to meet the needs of various industries while at the same time guaranteeing uniqueness for all EPC-compliant tags, called *EPC tags*. EPC encoding schemes typically contain a serial number, called an *EPC number,* which can be used to uniquely identify an object. The EPC number is a structured number composed of multiple fields, as shown in Table 5.5.

Table 5.5 Fields of an EPC Number

Field Name	Description	Example (Hexadecimal)
Header	Identifies the length, type, structure, version, and generation of EPC	015

Continued

Table 5.5 continued Fields of an EPC Number

Field Name	Description	Example (Hexadecimal)
EPC manager number	Identifies the company or company entity	35000
Object class	Identifies the product, similar to a stock keeping unit (SKU)	213761
Serial number	Identifies this item of this product: the specific instance of the product being tagged	210000000

EPCglobal Network-compliant software and hardware will use EPCglobal standard data protocols and therefore will use EPC Manager numbers. Hence the EPC Manager numbers issued by EPCglobal are required if companies will engage with trading partners outside their internal operations.

An example of an EPC number is shown in Figure 5.1. Additional fields may also be used as part of the EPC number to properly encode and decode information from different numbering systems into their native (human-readable) forms.

Figure 5.1 Structure of an EPC Number (Fields Are Explained in Table 5.5)

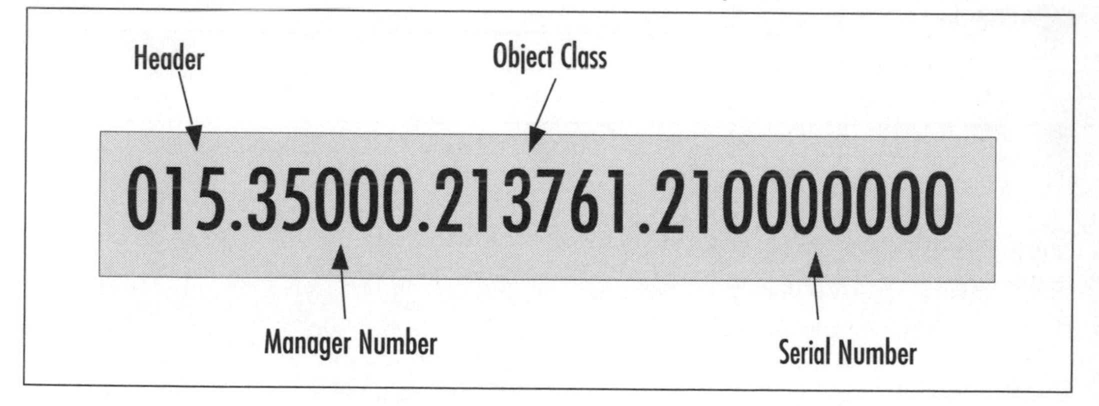

The EPC system also defines the tag classes, which are discussed in Chapter 4.

Of course, these regulations and standards have their impact on the RFID industry, including users. Most of the standards we have looked at so far can be categorized into two groups: tag data standards and air interface protocols.

Air Interface and Tag Data Standards

An RFID system consists of two main functionalities:

- **Tag data** Writing data to a tag and reading data from the tag.
- **Communication** Transferring data—for example, between a reader and a tag.

Corresponding to these two functionalities are two groups of standards: standards for tag data formats and standards for air interface protocols.

Tag Data Standards

Tag data standards are protocols that specify how to write data to a tag and how to read data from the tag. In other words, they specify the encoding, decoding, and formats of data. The whole does not need to be stored on a tag. You can store data about an item in a database and let the data fields on the tag point to this data. For example, after you retrieve the EPC number from a tag, this number can point to the data about the item stored in a database. Tag data formats can be used to accomplish the following:

- Specify the tag architecture.
- Identify a specific item—for example, during shipment.
- Specify the available memory size on a tag.

CAUTION

You must know the difference between air interface protocols and tag data standards.

Air Interface Protocols

Air interface protocols specify the rules for communication between a tag and a reader. This set of protocols include anticollision algorithms to deal with the dense environments as well as modulation and demodulation—that is, coding data into the outgoing carrier signal and decoding data from an incoming signal. Some features of an air interface protocol may depend on the operating frequency. For example, multiple standards with the name ISO/IEC 18000-*i*, where *i* is an integer, correspond to different frequencies.

Impact of Regulations and Standards

RFID regulations and standards have definitely made an impact on the RFID industry in various areas, including business operations and IT infrastructure. The impact has both advantages and disadvantages.

Advantages of Regulations

The regulations in the RFID area made by different countries and regulatory organizations have the following positive impacts:

- They lower the risk of adverse radiation effects.
- They pave the way for healthy market competition by regulating the areas where foul play could otherwise happen. For example, regulating the frequency and the maximum power emission secures the existing RF technologies and services from disruption and secures the public from adverse radiation effects. At the same time, it also forces vendors to compete in other areas such as features, price, and customer service.
- Regulations also help advancement of technology by directly or indirectly making it easier for more players to enter the market, thereby promoting entrepreneurship.

Advantages of Standards

What would there be without standards? Chaos: The same product from different vendors would work in different ways, and as a result, two instances of the same product from different vendors would not be interoperable with each other. In general, the following are the advantages of having standards in the RFID industry:

- All devices following the same standard will be interoperable with each other. This helps consumers and therefore vendors as well.
- Standards promote automation and thereby reduce duplication of effort. For example, whatever is standardized can be built once and used in other applications where it is needed rather than building it over and over again in the same or different ways.
- Because the standards bodies are not serving the interest of just one vendor, the standards generally define the most efficient platform on which an industry can operate and advance.
- Standards generally reduce cost and ease implementation.
- Standards develop consumer confidence in the technology.

Looking at the other side of the coin, regulations and standards do have their disadvantages.

Disadvantages of Regulations and Standards

The following are the disadvantages of regulations and standards:

- The highest limit on the emitted power sets the highest limit on the read range of a tag.
- Passive tags are especially affected by the highest limit on the emitted power because they depend on the reader for the power that they use to power up their circuitry and to compose the response signal.
- Because the regulated operating frequencies are different in different regions of the world, an RFID device that works in one region will not work in another region.

The advantages of regulations and standards often outweigh the disadvantages.

Regulatory and Standards Bodies

We have mentioned several organizations responsible for regulations and standards in the RFID industry. A list of these organizations is provided in Table 5.6.

Table 5.6 The Main Regulatory and Standards Bodies in RFID

Organization	Function
International Telecommunication Union (ITU)	An international organization established to standardize and regulate international radio and telecommunications; it organized the world into three regulatory regions for RFID
European Telecommunications Standards Institute (ESTI), created by the European Conference of Postal and Telecommunications (CEPT)	Regulates RFID in Europe
Federal Communications Commission (FCC)	Regulates RFID in the United States
Ministry of Public Management, Home Affairs, Posts and Telecommunications (MPHPT)	Regulates RFID in Japan
Office of the Telecommunications Authority (OFTA)	Regulates RFID in Hong Kong
Standardization Administration of China (SAC)	Issues regulations for RFID in China
EPCglobal	Develops standards for the EPCglobal network
International Organization for Standardization (ISO)	Develops standards for RFID and several other industries

The three most important takeaways from this chapter are the following:

- To allot the frequency usage for RFID devices (especially in the UHF band), the world is organized into three regulatory regions: Region 1 includes Europe and Africa, Region 2 includes the Americas, and Region 3 includes Asia and Australia.
- The two organizations that develop standards for RFID are the ISO and EPCglobal.
- The main advantage of RFID regulations is to make RFID devices safe and to prevent them from disrupting existing services in the RF arena. The main advantage of standards is the resulting interoperability.

Summary

The regulations in the RFID industry serve two main purposes: to keep RFID devices safe (for example, in the area of human exposure to radiation) and to prevent RFID devices from disrupting the existing services in the RF arena. These goals are established by regulating the maximum power emitted by the devices and by regulating operating frequencies. The world is organized into three regulatory regions: Region 1, which includes Europe; Region 2, which includes the Americas; and Region 3, which includes Asia and Australia.

The main purpose of RFID standards is to ensure interoperability: Different systems and components work together effectively. This helps an industry to advance, and it also helps create healthy marketing competition among vendors. There are two main standards organizations in the RFID arena: ISO, which develops standards for several industries, and EPCglobal, which is specific to RFID.

By now you have learned about tags, readers, the physics of RFID, and RFID standards and regulations. Equipped with this knowledge, you are prepared to select the design of your RFID system, which is the topic of the next chapter.

Exam's Eye View

Comprehend

- ☑ The read range of a reader decreases as a square root of the power emitted by its antenna.
- ☑ The electronic product code (EPC) is a family of coding schemes for Gen 2 tags.
- ☑ The EPC encoding schemes typically contain a serial number, called an *EPC number*, which can be used to uniquely identify an object.
- ☑ Air interface protocols are a set of protocols that define the rules for communication between tags and readers.

Look Out

- ☑ Of course, all interrogators can read tags, but some interrogators can also write to tags.
- ☑ The UHF RFID systems have evolved at different frequencies in different regions of the world.

- ☑ The former Soviet Union is included in regulatory Region 1 along with Europe, whereas Australia is included in Region 3, along with Asia.
- ☑ HF RFID devices use 13.56 MHz operating frequency all over the world.

Memorize

- ☑ EPCglobal is a joint venture between GS1 (formerly known as EAN International) and GS1 US (formerly the Uniform Code Council Inc.) and was created to commercialize EPC technology, which was originally developed at the Auto ID center at MIT.
- ☑ The two organizations that develop standards for RFID are the ISO and EPCglobal.

Key Terms

Air interface protocols The set of protocols that define the rules for communication between tags and readers.

Electronic Product Code (EPC) A family of coding schemes for Gen 2 tags.

EPCglobal A joint venture between GS1 (formerly known as EAN International) and GS1 US (formerly the Uniform Code Council Inc.), created to commercialize the EPC technology that was originally developed at the Auto ID center at MIT.

EPCglobal Network A set of RFID technologies that enables immediate automatic identification and sharing of information on items in the supply chain.

EPC number A serial number, a part of an EPC coding scheme, which can be used to uniquely identify an object.

Federal Communications Commission (FCC) An independent U.S. government agency established by the Communications Act of 1934 as the successor to the Federal Radio Commission and charged with regulating all nonfederal government use of the radio spectrum (including radio and television broadcasting) and all interstate telecommunications (wire, satellite and cable) as well as all international communications that originate or terminate in the United States.

International Electrotechnical Commission (IEC) The IEC is an international standards organization in the area of electrical, electronic, and related technologies. Some of its standards are developed jointly with the ISO.

Institute of Electrical and Electronics Engineers (IEEE) An international nonprofit, professional organization for the advancement of technology related to electricity and electronics. There are about 900 active IEEE standards.

Interoperability The ability of systems or components of a system to provide services to and accept services from other systems or components and thereby operate together effectively to provide services to the user.

International Organization for Standardization (ISO) An international standards body composed of representatives from national standards bodies. Founded on February 23, 1947, this organization sets worldwide industrial and commercial standards, which are popularly called *ISO standards*.

International Telecommunication Union (ITU) An international organization established to standardize and regulate international radio and telecommunications. It was originally founded with the name International Telegraph Union in Paris on May 17, 1865.

Regulation A legal restriction promulgated by a government administrative agency through rule making and typically supported by a threat of consequences such as fines for not following it.

Specific absorption rate (SAR) A measure of the rate of energy absorbed by (or dissipated in) an incremental mass contained in a volume element of dielectric materials such as biological tissues.

Standard Guideline documentation (specifications) that reflects agreements on products, practices, or operations by nationally or internationally recognized industrial, professional, or trade associations or governmental bodies. If all the vendors follow the same standard, the products from those vendors will be compatible with each other and will be interoperable.

Self Test

A Quick Answer Key follows the Self Test questions. For complete answers and explanations to the Self Test questions in this chapter as well as the other chapters in this book, see **Appendix A**.

1. The United States is included in which of the following regulatory regions for RFID?

   A. Region I

   B. Region 2

   C. Region 3

   D. Region 4

2. What is the most commonly used RFID frequency in the LF range?

   A. 13.56 MHz

   B. 200 KHz

   C. 125–134 KHz

   D. 125–134 MHz

3. What is the purpose for regulating the power emitted by the antenna of an RFID device?

   A. To avoid interference with RF waves from other devices

   B. To save energy

   C. To help the environment

   D. To avoid disrupting existing RF services

4. Air interface protocols define rules for all of the following except:

   A. Communication between a reader and a tag

   B. Bit encoding and decoding (modulation and demodulation)

   C. Commands for reading from and writing to tags

   D. Formatting EPC numbers

5. What kind of standards define the rules for communication between an interrogator and a tag?

A. Air interface protocols

B. Tag data format standards

C. RS-232

D. TCP/IP

6. Which organization came into existence to standardize and commercialize the RFID technology developed at the Auto-ID center at MIT?

A. ITU

B. ISO

C. EPCglobal

D. FCC

7. An EPC number typically contains all of the following fields except:

A. Header

B. Cyclic redundancy check (CRC)

C. Manager number

D. Serial number

8. Which of the following tag types is most affected by regulating the maximum radiated power?

A. Active

B. LF

C. Semipassive

D. Passive

9. All of the following are elements of the EPCglobal network except:

A. Electronic product code (EPC)

B. DHCP

C. ID system for tags

D. Object name service

10. Specific absorption rate is:

    A. Linearly proportional to the electric field strength of the wave being absorbed.

    B. Directly proportional to the square of the electric field strength of the wave being absorbed.

    C. Inversely proportional to the electric field strength of the wave being absorbed.

    D. Inversely proportional to the square of the electric field strength of the wave being absorbed.

Self Test Quick Answer Key

For complete answers and explanations to the Self Test questions in this chapter as well as the other chapters in this book, see **Appendix A**.

1. A
2. C
3. A
4. D
5. A
6. C
7. B
8. D
9. B
10. B

Chapter 6

RFID+

Selecting the RFID System Design

Exam Objectives	What It Really Means
5.1 Given a scenario, predict the performance of a given frequency and power (active/passive) as it relates to: read distance, write distance, tag response time, storage capacity.	You should know the frequency ranges available for RFID devices, and how performance metrics depend on the frequency. Understand that the only globally accepted radio frequency for RFID systems is 13.56 MHz, which falls in the HF band, and that the frequencies in the UHF range allocated to the RFID devices are different in different regions of the world. You must understand that the high frequency is not necessarily a good idea if not demanded by the application because high frequencies are more vulnerable to absorption.
5.2 Summarize how hardware selection affects performance (may use scenarios).	Understand the important performance related factors that need to be considered in selecting hardware components such as readers, tags, antennas, transmission line, and mount points for readers. You must understand how most of these factors are determined directly or indirectly by application requirements, operating conditions, and compliance with standards.

Introduction

You will design your RFID system to meet application performance requirements. Generally speaking, RFID is a means to identify an object by using radio frequency transmission, which suggests that communication is involved in the identification process. The communication takes place between a reader and a tag, which should be tuned to the same frequency. RFID systems are available at different frequencies. To select the right frequency for your system, you need to understand how the various performance parameters, such as read range, tag response time, and storage capacity, depend on the frequency. This understanding will also help you select the correct hardware components for your RFID system, such as readers, tags, and antennas. The tags are attached to the items that need to be identified and tracked, whereas readers will be mounted at places from where they will read the tags.

So, the core issue in this chapter is how to design your RFID system. To put our arms around this issue, we will explore three avenues: selecting operating frequency, selecting hardware components, and selecting mount points for readers.

Understanding RFID Frequency Ranges

A tag and a reader use radio waves of a certain frequency, called the *operating frequency*, to communicate with each other. Radio waves are electromagnetic waves that cover part of the electromagnetic spectrum of frequencies, called the *radio frequency spectrum*. Because RFID systems generate and radiate the electromagnetic waves that fall along the radio frequency spectrum, they are justifiably classified as radio systems and are regulated as such. However, other radio services have operated before the arrival of RFID systems. Radio, television, mobile radio services (police, security services, and industry), marine and aeronautical radio services, and mobile telephones are just a few. Therefore, it is important to ensure that these services are not disrupted or impaired by RFID systems. This requirement significantly restricts the suitable operating frequency ranges for RFID systems. For this reason, the so-called industrial, scientific, and medical (ISM) frequencies, originally reserved for noncommercial uses in industrial, scientific, and medical fields, are used for RFID systems.

Table 6.1 shows the radio frequency ranges that are of interest to RFID systems, along with the ISM frequencies. RFID systems use many different frequencies in the radio frequency spectrum, but there are four most commonly used radio frequency ranges: low frequency (30–300 KHz), high frequency (3–30 MHz), ultra-high frequency (300 MHz–3 GHz), and microwave frequencies (1–300 GHz).

Table 6.1 Radiofrequency Ranges in Which RFID Systems Can Operate and Read Distance by Frequency

Name	Frequency Range	Wavelength Range	ISM Frequencies	Read Range For Passive Tags
Low frequency (LF)	30 KHz–300 kHz	10 km–1 km	125–135 KHz	<50 cm
High frequency (HF)	3–30 MHz	100 m–10 m	6.78 MHz, 8.11 MHz, 13.56 MHz, 27.12 MHz	<3 m
Ultrahigh frequency (UHF)	300 MHz–3GHz	1 m–10 cm	433 MHz, 869 MHz, 915 MHz	<9 m
Microwave frequency	1–300 GHz	30 cm–1 mm	2.44 GHz, 5.80 GHz	>10 m

Table 6.1 also shows the read range for passive tags corresponding to each frequency range. Active tags can have a read range of up to 100 meters. For example, active tags used on large assets such as cargo containers, rail cars, and large reusable containers, which usually operate at 455 MHz, 2.45 GHz, or 5.8 GHz, typically have a read range of 20 meters to 100 meters.

NOTE

Because LF RFID systems operate over short distances, interference with the surroundings is less an issue. This results in the system's increased accuracy and security.

As shown in Table 6.2, regulatory bodies have chosen different ranges for RFID within the UHF band in different parts of the world. Broadly speaking, most of the countries have allocated the RFID bands from the following three ranges:

- **Range 1: 865-868 MHz** For example, the bands allocated in India and Europe fall in this range.
- **Range 2: 902-928 MHz** For example, the bands allocated in the United States and Australia fall in this range.
- **Range 3: 950-954 MHz** For example, the bands allocated in Japan fall in this range.

Table 6.2 UHF Bands Allocated for RFID Systems Worldwide

Area	UHF Frequency Band Allocated to RFID Systems	Maximum Power Emission
United States	902–928 MHz	4 W (EIRP)
Australia	918–926 MHz	1 W (ERP)
Europe	865–868 MHz	2 W (ERP)
Hong Kong	865–868 MHz 920–925 MHz	2 W (ERP) 4 W (EIRP)
India	865–867 MHz	4 W (EIRP)
Japan	950–956 MHz	4 W (EIRP)
Singapore	923–925 MHz	2 W (ERP)

Note that in Table 6.2, the permitted radiated power, expressed in units of watts, is presented in different quantities. For example, in the United States, the radiated power is presented by EIRP, whereas Europe tends to use ERP. As demonstrated in Exercise 6.1, the conversion must be done between ERP and EIRP when necessary while comparing these numbers.

CAUTION

Today, the only globally accepted radio frequency for RFID systems is 13.56 MHz, which falls in the HF band.

EXERCISE 6.1

Calculate the ratio of the maximum allowed power emission by an RFID system in the United States and Europe.

Solution:

The power allowed in the United States = 4W EIRP

The power allowed in Europe = 2 W ERP = 1.64 x 2 W EIRP = 3.28 EIRP

Ratio = U.S./Europe = 4.00 W/3.28 W = 1.22

So, RFID systems operate in four main ranges of the radio frequency spectrum: LF, HF, UHF, and microwave. Although the choice of frequency does not affect the underlying physics of how the system components will operate, it does affect the system's performance in areas such as speed, range, and accuracy.

RFID Frequency Ranges and Performance

While designing your RFID system, you will need to decide at which frequencies the RFID devices (readers and tags) will operate. To decide wisely, you need to know the applications' performance requirements and how your frequency choices will impact performance. Let's take a close look at the way frequency ranges affect various performance metrics.

The Low-Frequency (LF) Range

The LF range extends from 30 KHz to 300 KHz. The RFID systems in this range typically operate at ISM frequencies 125 KHz and 134 KHz. Some important characteristics related to performance of RFID systems operating in the LF range are as follows:

- **Short read range** The read range of RFID systems operating in the LF range is short: less than half a meter.
- **Lower reading speed** In general, the higher the frequency, the longer the read range, and the higher the data transfer rate will be. Data transfer rate is directly proportional to available bandwidth.
- **Less absorption** Because wavelength is inversely proportional to frequency, the lower the frequency of an RFID system, the higher is the wavelength. Due to the higher wavelength, the LF signals are not easily absorbed by the atmosphere and the material they move through. For this reason, RFID systems operating in the LF range work well around water and metal.

Due to the short read range and less absorption, LF systems are more robust to external influences. Based on these characteristics, the following are common applications of RFID systems in the LF range:

- Access control
- Animal and personnel tracking
- Vehicle immobilizers

TIP

The bandwidth available at low frequency is very limited, which results in very slow data transfer rates. For example, in the case of the International Standard ISO 18000 Part 2 covering LF RFID systems, the command signaling rate (meaning the communication speed between reader and tag) is only around 5 kbits/second.

The next step on the frequency ladder is HF.

The High-Frequency (HF) Range

The HF range extends from 3 MHz to 30 MHz. RFID systems in this range usually operate at 13.56 MHz, which is a globally accepted frequency for RFID systems. Some important performance-related characteristics of RFID systems operating in this range are described in the following:

- The read range is about 3 m.
- Due to shorter wavelengths, the signals in this range cannot penetrate through materials as well as the LF signals can.
- This frequency range provides greater options in data transfer speed, compared to LF.

Due to these characteristics, following are typical applications for RFID systems in the HF range:

- Building access control
- Item-level tracking, including baggage handling
- Libraries

Because 13.56 MHz is the globally accepted frequency standard for RFID systems, the HF RFID systems have been more broadly adopted.

The next step on the frequency ladder is UHF.

Ultra High Frequency (UHF) Range

The UHF range extends from 300 MHz to 3 GHz. As Tables 6.1 and 6.2 show, the actual frequencies being used by RFID systems operating in this range are 344 MHz and 860–960 MHz. The reading speed and data transfer rate for these systems can be high. However, systems in this range are relatively new and encounter a host of problems, some of which are described in the following:

- Due to smaller wavelength, the RF energy can be easily absorbed by liquids and matter. It can considerably reduce the reading range.
- The high reading speed creates more probability for errors.
- As shown in Table 6.2, countries have allocated different frequencies for RFID systems in this range; therefore, a UHF system that works in one country might not work in another country.
- Many consumer devices also operate in the same frequency range; therefore, RFID systems in this range are subject to interference with their signals.

The higher reading speed and longer read distance make these systems attractive for the following applications:

- Automated toll collection
- Warehouse management
- Inventory tracking

The next step on the frequency ladder is the microwave frequency range.

The Microwave Range

The microwave range extends from 1 GHz to 300 GHz. RFID systems in this range operate at ISM frequencies, 2.44 GHz and 5.80 GHz, which offer high data transfer rate. Following are some of the characteristics of microwave RFID systems:

- High reading speed and data transfer rate
- Long read distance
- Poor performance around water and metal

Due to these characteristics, the microwave RFID systems are used in the following applications:

- Long-range access control for vehicles
- Vehicle identification
- Automated toll collection
- Supply chain

Advantages and disadvantages of various frequency ranges and the typical RFID applications corresponding to each range are summarized in Table 6.3. Figure 6.1 shows how some characteristics of an RFID system depend on the frequency.

Table 6.3 Characteristics of Various Radiofrequency Ranges

Frequency Band	Advantages	Disadvantages	Typical RFID Applications
LF	Can work well around water and metal; accepted worldwide	Short read range and slow read speed	Animal identification, product authorization, close read of items with high water content
HF	Better accuracy and read speed, easier to read at a distance, can carry more information	Requires higher power	Building access control, airline baggage, libraries
UHF	Faster read speed, easier to read at a distance, can carry more information	Does not work well near water or metals	Parking lot access, automated toll collection, supply chain
Microwave	Faster read speed	Does not work well near water or metals	Vehicle identification, automated toll collection, supply chain

Figure 6.1 The Dependence of Some RFID System Characteristics on Frequency

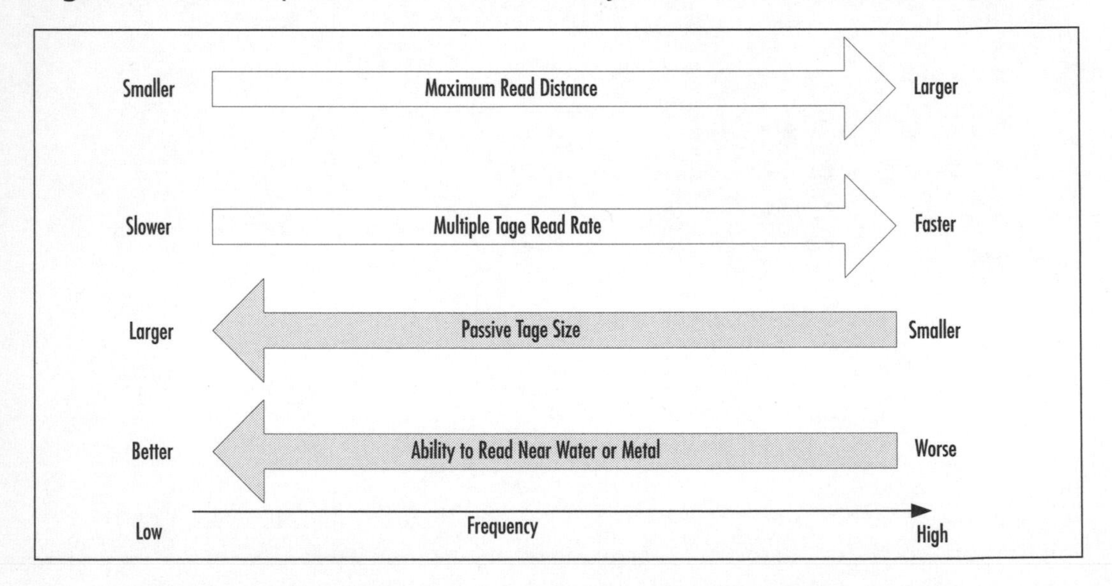

In a nutshell, RFID tags and readers must be tuned to the same frequency to communicate with each other. Frequencies exhibit different behavior in terms of characteristics that make them more useful for different applications. For example, LF tags are cheaper than UHF tags, use less power, and are better able to penetrate through water. Therefore, they are suitable for scanning objects with high-water content, such as fruit, at close range. UHF systems typically offer better read range and speed and can transfer data faster. But they use more power and are less likely to pass through materials. Combine this information with the fact that UHF waves tend to be more "directed," they require a clear path between the tag and reader. These characteristics make UHF systems more suitable for applications such as scanning boxes of goods as they pass through a bay door into a warehouse.

The read range performance of an RFID system is an important characteristic and is typically determined to a large extent by the following factors:

- **The power radiated by the reader, which is regulated** Passive tags get their power from the energy coming from the power radiated by the reader.
- **The operating frequency** The power received by the tag's receiver depends on the antenna size, which in turn depends on the frequency (or wavelength) of the received signal.
- **The sensitivity of the tag** The maximum read distance depends on the tag's power requirements, which in turn depend on the tag type (active or passive) and the tag's antenna size.
- **Communication technique** The read range also depends on which communication technique the RFID system is using: inductive coupling or backscattering.

Now that you understand how frequency choices impact RFID system performance, you are well equipped to start exploring ways to select operating frequency for your RFID system.

Selecting Operating Frequency

By now you know that RFID devices (tags and readers) are available at different frequencies. You also know how different performance metrics depend on the frequency. But how do you decide at which of these available frequencies your system should operate? The short answer to this question is that it all depends on the application requirements and the operating conditions. That said, following are the main factors that you should consider in selecting operating frequency:

- **Application types** Because all applications in each application type such as retail, automatic toll collection, and animal tracking have a common set of requirements, most application types are associated with specific frequencies.

- **Read range** Read range depends on frequency, among other factors. So, the read range requirement of your application will give you a very good idea as to which frequency you should select for your system.
- **Operating conditions** In making a frequency selection, you should also factor in the conditions under which your system will operate. For example, if there is water (or water-related conditions, such as mud or snow) or metal in the vicinity of the RFID system, LF and HF are the ideal frequency selections. This is because LF and HF can penetrate through these materials better than UHF and microwave frequencies can.

Now that you understand how frequency choices impact RFID system performance and how you select the operating frequency, you are well equipped to start exploring how to select individual components of an RFID system, such as tags.

Selecting Tags

To select the right tags for your application, you need to consider various factors such as tag kind (tag types and tag classes), operating frequency, read range, data capacity, tag form and size, environmental conditions under which the tags will operate, and the standards and regulations with which you need to comply. Most of these characteristics have already been discussed in Chapter 4 and in this chapter. Here we present a brief discussion regarding the role of these factors in selecting tags.

Kinds of Tag

The kind (type and class) of tag that you select depends on the application requirements. Following are some examples and scenarios.

- **Tag Types** If the application simply requires the tag to store some data such as identification number and provide it on request, you can use passive tags. However, if real-time features such as sensing the temperature and humidity are required, you must select active tags, because your data will need real-time processing.
- **Tag Classes** Tag classes offer different features such as read only (RO), write once and read many (WORM), and read and write (RW). If the application requires the tag to store a unique identifier that will not change and provide it when requested, simply use RO or WORM tags. If the application requires the tag to store dynamic data (data that is subject to change), you need RW tags.

Operating Frequency

This chapter has already discussed the frequencies available for RFID systems and how to make the frequency selection. There are two important points to note:

- Different frequency bands are allotted to RFID systems in different regions of the world.
- From the available frequencies, you need to select the right frequency for your system based on the application requirements. The higher the frequency, the larger the read range.

Read Performance

Read performance of a tag depends on the following factors:

- **Read range** The read range is the maximum distance from which a reader antenna can read a tag. This range is required by the application for which you are selecting the tag. Read range is discussed in detail in Chapter 4.
- **Antenna polarization and orientation** Antenna polarization and tag orientation, discussed in Chapter 4 and further in this chapter, also affect read performance. For optimal reading, tag orientation should be consistent with antenna polarization.
- **Reading efficiency** The reading efficiency, also called *read robustness*, is the ratio of the number of successful reads to the total number of read attempts. This is the ultimate factor that needs to be optimized to improve performance.

CAUTION

It's very easy to get carried away in maximizing the read range. But it's really the reading efficiency that you should be maximizing. The read range should only satisfy the application requirement. Unnecessarily high read ranges may have a negative impact on the system in terms of interference and security.

The factors that can cause a reduction in reading efficiency include the following:

- *Attenuation*, a decrease in the signal amplitude, caused by different product and packaging materials (such as liquids and metals) when the signal passes through them

- Presence of metal close to RFID antennas, or a large (as compared to antenna size) mass of metal passing an antenna, which can create a mismatch between the characteristics of the antenna and the reader
- *Radio frequency interference (RFI)* from RF transmitters and electrical drives, motors, and power supplies in the location of the RFID antenna system

Data Capacity

The term *data capacity* refers to the amount of data (information) that can be stored in a tag. Increased data capacity increases the usefulness of the tag and its cost. While selecting the data capacity, you should consider the following factors:

- **Data amount** For applications that only require the tags to store the identification number, you can simply use tags that offer minimal storage, such as class 0 tags. For applications that require more data capacity, tags with appropriate memory can be selected accordingly.
- **Data security** Some applications could require data locking to prevent tempering with the tag data. Data locking can be implemented at either the hardware or software level. For read-only tags, such as class 0 tags, the identifier is permanently burned into the tag and cannot be changed. WORM and RW tags can use software locks by implementing password schemes.

Tag Form and Size

The tag form and size should be compatible with the item and the environment. For example, the tag needs to fit on the item. Tag forms and sizes are discussed in detail in Chapter 4. Following are the two main factors that you should consider regarding the tag form and size:

- **Tag dimensions** The dimensions of a tag should be suitable for the size and shape of the item that needs to be tagged. For example, you should consider the space available for the tag on the item without obstructing any critical information printed on the product's surface.
- **Tag ruggedness** Your application could require a rugged tag to withstand harsh environmental conditions such as corrosive chemicals, extremely high or low temperature, humidity, and mechanical shocks. A tag can be made rugged by enclosing it in a cover. Rugged tags are usually expensive.

CAUTION

Do not fold a tag to reduce its size. Folding it can detune the tag antenna, which then can fail to receive enough power from the reader antenna and therefore will fail to respond to the reader.

Environmental Conditions

Environmental conditions can affect the performance of a tag and therefore the selection of the tag for an application. You should consider the following environmental factors in selecting tags:

- Other objects in the neighborhood of the item to be tagged
- Other environmental conditions, such as extreme temperature and humidity

Standards Compliance

You should make sure that the tags you select meet the established and emerging standards. This is important to ensure compatibility and interoperability with other systems meeting those standards.

The two main components of an RFID system are tag and reader, and the two should be compatible with each other. So, the process of selecting tags is tied into selecting readers.

Selecting Readers

A reader's job is to collect data from tags and possibly send it to an application running on a host computer. This section discusses the factors that need to be considered in selecting readers.

Reader Types

When selecting a reader type, consider the following characteristics:

- **Operating frequency** This chapter has already discussed the frequencies available for RFID systems. There are two important points to note:
  - Different frequency bands are allotted to RFID systems in different regions of the world.
  - From the available frequencies, you need to select the right frequency for your system based on application requirements. The higher the frequency, the larger the read range.

- **Number of antennas** Most readers support a minimum of two and a maximum of four antenna ports. However, a reader can have one, four, or eight antenna ports. You cannot go wrong in simply choosing a reader with four antenna ports because it offers better flexibility in covering a wide read zone. The number of ports you need with a reader really depends on the application's needs.
- **Reader interfaces** Often the reader needs to send the collected data somewhere. So the readers come with I/O controllers which support the interfaces for sending out the collected data. Depending on the type of interface, it may connect the reader to a host computer serially, or it may connect the reader to a network.
- **The reader mobility** Based on application requirements, you might need a fixed or a mobile reader. Mobile readers are usually wireless readers. That is, they connect to the network using wireless technology.

Ability to Upgrade

The ability to upgrade the readers can considerably reduce the system cost in the long run. This ability should allow you to upgrade the firmware and fix bugs in it.

Installation Issues

While selecting a reader, you should also consider the installation requirements. Following are some examples:

- Properly installing a specific reader could require additional structure, such as a portal that needs to be built.
- The reader and its cable (transmission line) must not pose a risk to operations personnel in the area.
- The long cables to connect a reader to its antenna can attenuate the signals. If the distance between the antenna and the reader is such that the installation requires a cable longer than 6 feet, the cable better be of high quality and low loss.

Legal Requirements

The maximum power that transmitter can emit is regulated in most countries. It means that you must make sure that the selected readers comply with those regulations. It also means that you should not tamper with the features of the readers that comply with the regulations.

Manageability

Depending on the application requirements, you might need a reader that can be managed remotely, for example, using Simple Network Management Protocol (SNMP). This gives you the advantage of tracking, diagnosing, and fixing errors remotely, without manually visiting the site.

Quantity

You will also need to determine how many readers you will need for your application. This number will depend on the following:

- Number of read zones
- Number of reader antennas required for each zone
- Number of antenna portals on each reader

Ruggedness

Your application could require rugged packaging to withstand harsh environmental conditions such as corrosive chemicals, extremely high or low temperature, humidity, and mechanical shocks. A reader can be made rugged by enclosing it in a cover. Rugged tags are usually expensive.

By now, you have a very good idea of how to select tags and readers. Both tags and readers have antennas, which we'll examine next.

Working with Antennas

In general, an antenna is any structure or device used to receive or radiate electromagnetic waves. As you already know, both tags and readers have their own antennas. You need to select appropriate antennas for your RFID system because they come in various types and configurations. Before selecting an antenna, you should understand the antenna types.

Understanding Antenna Types

This section discusses the common antenna types listed in the following:

- Monopole antenna
- Dipole antenna
- Linearly polarized antenna
- Circularly polarized antenna
- Omnidirectional antenna

- Helical antenna

Before we could talk about antenna types, you should know the definitions of the following terms:

- **Channel** A single path provided by a transmission medium. This path may be provided by a cable or by a specific frequency.
- **Source** An object that encodes the message data and transmits it via a channel to one or more receivers.
- **Driven element** The single antenna that has an applied source feed.
- **Ground plane** An electrically conductive surface that serves as the reflection point near an antenna or as a reference ground in a circuit.

Dipole Antennas

A *dipole antenna* is an antenna with a center-fed driven element for transmitting or receiving radio frequency energy. From a physics viewpoint, this type of antenna is the simplest practical antenna. It consists of a straight electric conductor, made of conducting metal such as copper, interrupted at the center, therefore making two poles. As shown in Figure 6.2, the category of dipole antennas can be further subdivided as into the following:

- **Half-wavelength dipole** The total length of this antenna is half the wavelength corresponding to the frequency to be used. It optimizes the transfer of power between the tag and the reader.
- **Quarter-wavelength dipole** The total length of this antenna is a quarter the wavelength corresponding to the frequency to be used. It uses the reflective ground plane that provides an image of the antenna to complete the dipole.
- **Dual dipole antenna** As the name suggests, a dual dipole antenna consists of two dipoles. It covers more area and therefore reduces the sensitivity of a tag's orientation.
- **Folded dipole antenna** This antenna consists of two or more straight electric conductors that are connected in parallel, and each electric conductor is half the wavelength corresponding to the frequency to be used.

Figure 6.2 Various Kinds of Dipole Antenna

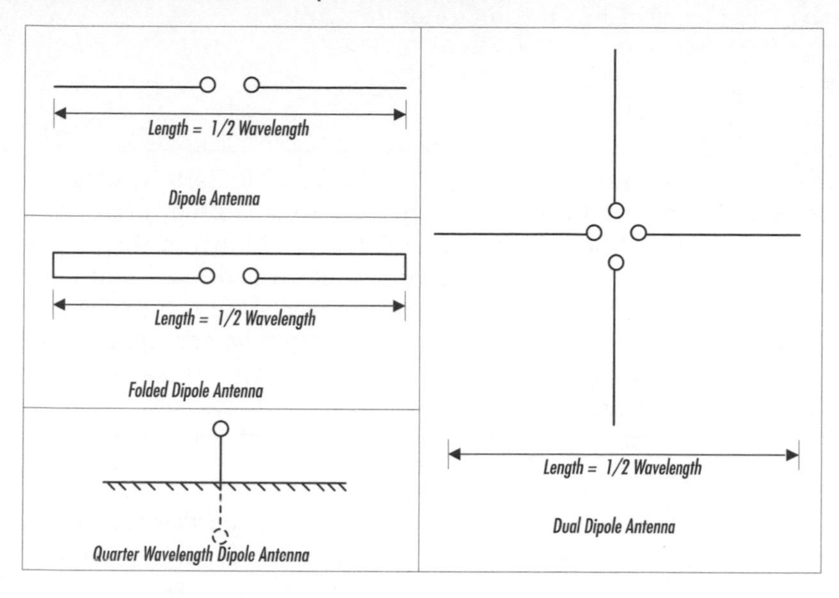

Monopole Antennas

A *monopole antenna* is a type of dipole antenna formed by replacing one half of the dipole antenna with the ground plane at a right angle to the remaining half. If the ground plane is large enough, the monopole behaves exactly like a dipole because its reflection in the ground plane forms the missing half of the dipole. The most common example of a monopole antenna is a whip antenna, which is basically a stiff but flexible wire, usually mounted vertically.

Linearly Polarized Antenna

As you know, as a wave travels, there are variations (or vibrations) in the wave, such as variations in electric field or magnetic field of an EM wave. As described in Chapter 1, the variations of electric and magnetic fields (vectors) in an EM wave are in a plane perpendicular to the direction of propagation of the wave. If the variations are such that the electric field vector stays parallel to a line in space as the wave travels, the wave is said to be linearly polarized, and the antenna that transmits such a wave is called a linearly polarized antenna. Because magnetic field (perpendicular to electric field) will also stay parallel to a line in space, this fixes the plane of the electric and magnetic field vectors. Therefore, linear polarization is also called plane polarization. The following two kinds of linear polarization are of special interest:

- **Horizontal polarization** This is the linear polarization in which the wave travels horizontal to the surface of the Earth.

- **Vertical polarization** This is the linear polarization in which the wave travels perpendicular to the surface of the Earth.

NOTE

Horizontally polarized waves (signals) travel parallel to Earth's surface, whereas perpendicularly polarized waves travel perpendicular to the surface of Earth. If you look carefully at TV antennas, you will find that most of them have rods about a meter in length set horizontal to the earth's surface. That means that the carrier waves for TV are polarized horizontally—that is, the electric vector is horizontal, the magnetic vector is vertical, and both lie in a plane perpendicular to the direction of propagation of the wave. So, antennas are set broadside to the direction of propagation of the wave.

A linearly polarized antenna emits a narrow radiation beam that increases the read range of a tag. However, a linearly polarized antenna of a reader is sensitive to the tag orientation with respect to polarization. Therefore, this type of antenna is useful for applications in which the tag orientation is fixed and known (predictable). Dipole antennas are linearly polarized.

Circularly Polarized Antennas

A traveling EM wave is said to be circularly polarized if the electric field vector rotates in a circle as the wave travels. The antenna that emits circularly polarized waves is called a *circularly polarized antenna*. A circularly polarized signal contains horizontal and vertical components. Therefore, a circularly polarized reader antenna is largely unaffected by tag orientation. For example, if the tag is oriented to receive horizontally polarized waves and the reader antenna is emitting circularly polarized waves, the tag will still receive the horizontal component of the signal power. For this reason, a circularly polarized reader antenna is preferred in applications in which the tag orientation is unknown or unpredictable.

Omnidirectional Antennas

An *omnidirectional antenna* is a nondirectional antenna that radiates power uniformly in all directions. An ideally perfect omnidirectional antenna is also called an *isotropic antenna*, which is really a theoretical antenna used as a reference to calculate quantities such as antenna gain and effective radiated power (ERP). Practically speaking, the antennas can provide omnidirectionality in one plane, such as in a horizontal plane—that is, the plane parallel to the surface of the Earth.

Helical Antennas

As shown in Figure 6.3, a *helical antenna* is an antenna that consists of a conducting wire wound in the form of a helix. A helical antenna is an example of a circularly polarized antenna. Note the following about helical antennas:

- The length of the antenna coil determines the antenna gain.
- The diameter of the antenna coil determines its wavelength.
- Because helical antennas are circularly polarized antennas, they can receive signals with any type of polarization, such as linear, horizontally linear, or vertically linear.
- A helical antenna can be clockwise polarized or anticlockwise polarized. Clockwise polarized antennas will have poor antenna gain when receiving a signal that is anticlockwise polarized.

Figure 6.3 Helical Antennas

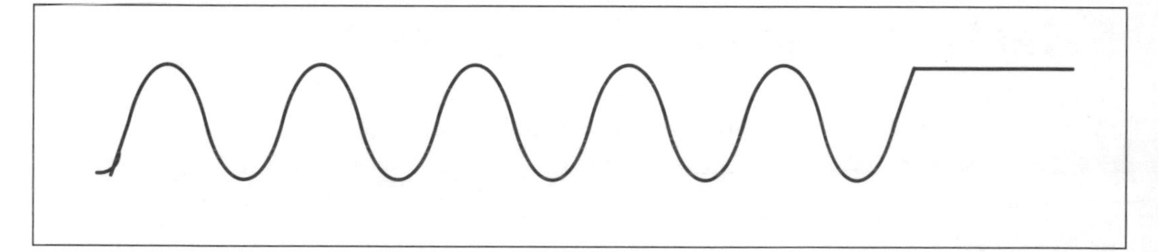

These antennas are best suited for applications such as animal tracking and space communication, where the orientation of the sender and receiver cannot be easily controlled or where the signal's polarization may change over time.

Selecting Antennas

When selecting antennas, you should consider the following factors:

- **Footprints** The footprint of an antenna is the ground area over which the antenna delivers a specified amount of signal power under specified conditions. That means a tag placed within the footprint of a reader's antenna can be read by the antenna. Because RFID antennas are mostly directional, the footprint of an antenna is rarely symmetrical around the antenna. You can use a device such as a spectrum analyzer to determine the actual footprint map of an antenna.
- **Polarization** Remember the following two things about polarization:

- If the tag orientation is arbitrary, unknown, or unpredictable, use circularly polarized antennas, because circularly polarized waves have both horizontal and vertical components. That means that if the antennas are circularly polarized, there will always be a transfer of some amount of power between the reader and the tag antennas, regardless of the tag orientation.
- If the tag orientation with respect to the antennas is known, use a linear antenna to receive the maximum power and thereby increase the read range.

- **Standards and regulations** You should be aware of the standards and regulations regarding the characteristics of RFID systems in your region of the world, such as the allowed operating frequency and the allowed maximum power to be emitted by the antenna. This is important for two reasons: to obey the law of the land and to be compatible with the environment.

The source of a reader that generates the signal is connected to the antenna through a transmission line.

Selecting Transmission Lines

In an RFID system, a transmission line is a physical medium (say a cable) used to connect the signal source to the antenna. The optimal transmission line would be the one that transfers the energy (power) from the source to the antenna with minimum power loss. When selecting a transmission line, you should be aware of the characteristics discussed in this section.

Impedance

You learned in Chapter 2 what impedance is and how an impedance mismatch between antenna and transmission line will create a reflected wave, which will result in decreased system efficiency. For optimal results, you must match the input impedance of the antenna with the characteristic impedance of the transmission line.

Cable Length and Loss

When you're choosing the physical length of a transmission line, keep in mind its electrical length, which is its length expressed as a multiple (or submultiple) of the wavelength of the signal that will propagate through it. Consider a cable that will transmit a signal of 3 GHz. The wavelength corresponding to 3 GHz is 10 cm. Now consider a 4 m cable. If the electrical length of the cable is expressed in units of l/4, the electrical length of this cable = 400/(10/4) = 160 units. So, the transmission line in this case is electrically too long. The longer the line, the greater will be the power loss.

Now consider a signal of 30 KHz propagating through the same transmission line. As shown in Exercise 6.2, this line is electrically too short for this signal.

EXERCISE 6.2

Calculate the electrical length of a 4 m cable in units of ?/4. The signal transmits a 30 KHz signal.

Solution:

$\lambda c/f = (3x10^8 \text{ m/s}) / (30000 \text{ 1/s}) = 10,000 \text{ m}$

So, the electrical length of the 4 m cable in units of quarter of a wavelength will be 4 / (10,000 / 4) = 0.0016. Therefore, this cable is too short for a signal of this frequency.

Transmission Line Types

The cable types most commonly used to form a transmission line are the following:

- **Coaxial cable** This cable consists of two coaxial conductors separated by a plastic insulating material. The inner conductor is a copper wire surrounded by the outer conductor, which is a braided wire jacket, a copper mesh. The outer conductor is then shielded with an insulating material. This cable type is useful to transmit low-amplitude signals because it can protect (shield) the signal from external interference. This is because the electromagnetic field carrying the signal exists only in the space between the outer and the inner signal. So the signal is shielded from external interference, which results in low loss. Following are the advantages of this cable type:
    - Low loss
    - Can be used to efficiently transmit low-amplitude signals
    - Useful for frequencies up to 3 GHz
- **Shielded pair cable** This cable consists of two parallel conducting wires embedded in solid dielectric material, which is surrounded by braided copper tubing, which in turn is surrounded by a rubber cover. The braided copper tubing acts as an electrical shield against external electromagnetic interference. The rubber cover protects the line from mechanical damage and moisture.

So, a reader has a source, a transmission line, and an antenna. The reader needs to be mounted at a place from where it will read the tags.

Mounting Equipment for RFID Systems

Tags attached to items carry information about the items. These tags need to be read by readers to retrieve the information. An *RFID portal* is the area where RFID tags can be read or written to. The portals can be grouped into two categories:

- **Stationary portals** These are the portals on which readers are mounted at a predetermined fixed place and wait for the tags to pass through their interrogation zone. This kind of portal is used in applications in which the path of the items containing tags is predetermined, such as along a conveyor.
- **Mobile portals** These are the portals in which readers are moved around to read the tags. Such portals are useful for applications in which the tagged items do not travel a predetermined path.

The common candidate portal points for mounting your RFID system are the following:

- Conveyer
- Dock door
- Forklift
- Point of sale
- Smart shelf
- Stretch wrap station

These portal points are discussed in the following sections.

Conveyors

Conveyors are used for case-level tracking—for example, in airports. To achieve the optimal results, multiple reader antennas should be used. Reader antennas are often mounted on gantries placed around the conveyor, as shown in Figure 6.4. The reader antennas on each side of the gantry will cover four faces of the container.

Figure 6.4 The Front View of a Conveyor Portal

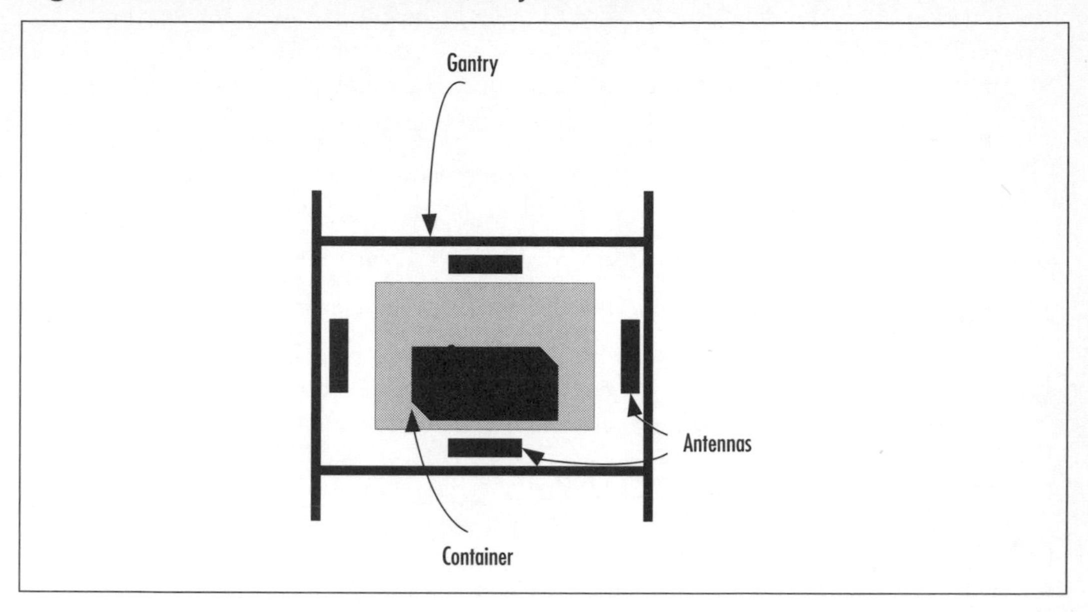

For optimal reading, consider the following factors in selecting a conveyor:

- The conveyor belt and the roller for the conveyor at the read point should be made of RF-friendly material, not of metal.
- The speed of the conveyor belt should adjusted for optimal tag reading.

Conveyors are good for case-level reading, whereas dock doors are suitable for pallet-level reading.

Dock Doors

A *dock* is a platform on which trucks or trains load or unload cargo. Keep in mind that the portal readers on a dock door might have to work in the presence of other electronic devices such as detectors and RF-reflective surfaces such as metal mesh. For example, the metal mesh surrounding the doorway could prevent reading of the tags going through adjoining doors.

In general, a door portal does not necessarily have to have a physical door. The term *door portal,* in general, refers to a vehicle carrying items in transit from one area to a different area. For example, when a truck parks and unloads at a dock door, items are stored or retrieved through a causeway, and a vehicle (mobile equipment) passes through the interior of racking aisles. Depending on the nature of the door portal (application), antennas can be mounted in various configurations. For example, multiple antennas are used in an array on both sides of the door to form the interrogation zone. When a transport vehicle such as a cart, a clamp truck, or a hand pallet truck passes through this interrogation zone, the door portal will read the tags attached to the items on the vehicle.

In setting up a door portal, you should consider the following:

- Your configuration of reader antennas should be able to cover (in reading) an area about 3 meters high and about 3 meters wide.
- The antennas need to be arranged in a sequence (that is, in an array) on each side of the door to form an effective interrogation zone.
- It is important to meet the minimum effective power level across the surface of the interrogation zone.
- In the case of a motion-triggered portal, the readers must be turned on in a timely fashion so that they can read in the minimum effective duration.

An alternative to a door portal is a forklift.

Forklifts

A *forklift* is a powered industrial truck used to lift and transport loads of materials by means of steel forks inserted under the load. A forklift is most commonly used to move a load stored on a pallet. These are especially well suited for reading tags from items on a pallet due their mobility and flexibility. The forklift can be connected to vehicle-mounted, data-collecting computers to inventory items efficiently. Following, however, are the disadvantages of mounting antennas on a forklift:

- A forklift-based reading system requires manual intervention from an operator.
- The communication devices used by forklift operators can cause RF interference and affect the readability of tags.
- Metallic forks can reflect RF signals from the RFID system and therefore could prevent reading of some tags.
- The speed of the forklift can also affect readability.

Stretch Wrap Stations

A *stretch wrap* has containers sitting on a turntable, which continuously changes the location and orientation of the tags. This is what makes a wrap station an attractive portal for reader antennas. Because of the turntable, the reader antenna has two advantages:

- As the pallet spins, it can make multiple attempts to read a tag.
- It gets opportunities to read tags in various orientations.

Because a stretch wrap station is usually the final step before shipping, an RFID system at this place guarantees the integrity of the containers.

Item-level tracking can be done at the point of sale.

Point-of-Sale Systems

An RFID *point-of-sale (POS) system* consists of scanning and payment capabilities. The RFID scanning technology used in POS systems can scan a whole shopping cart of goods or a basket full of items, say grocery items, in a few seconds. Here is how it works: As a customer pushes her cart to a designated checkout area, the readers mounted in the area collect the information about the products and their quantity in the cart. The readers send the information to the payment system, which displays the amount to be paid. The customer, at this point, can cancel the transaction or make payment through a credit or debit card. This payment transaction feature is built into the POS system.

Following are the benefits offered by the RFID POS system:

- Cost saving by eliminating the need for cash counter operators
- All stock-related store records can be updated in real time by taking feed from the RFID system.
- The possibility of shoplifting is reduced because all items going out of the shop can be tracked by appropriately setting up the checkout areas.
- Stock can be replenished efficiently by examining stock-out reports any time.

In addition to point of sale, retailers might also need to keep track of items on the shelves.

Smart Shelf

A *smart shelf* is a shelf that has readers mounted on it to read tags on the items on the shelf. When a customer picks up a tagged item from the shelf, the reader can no longer read the tag of this item, and this information flows to the inventory system, which assumes that the item has been removed from the shelf. Depending on its features and configuration, the inventory system might take further action, such as notifying store personnel to put more items on the shelf to avoid an out-of-stock situation. Here are some advantages of a smart-shelf system:

- Efficiently notifies store personnel of misplaced items
- Helps reduce out-of-stock situations
- Helps maintain better efficiency in inventory management
- Helps determine the sale potential of an item in a timely fashion

Looking at the other side of the coin, following are the disadvantages of a smart-shelf system:

- Because a smart-shelf system uses stationary readers and tracks individual items, lots of readers and tags would be needed. Therefore, cost becomes a significant factor to consider.
- Multiple reader antennas required in the same shelf could introduce overlapping interrogation zones and therefore interference of signals.
- If the items are densely packed on a shelf, stationary readers could miss some items, resulting in inventory issues.

The three most important takeaways from this chapter are the following:

- Read range and read speed increase with increase in frequency, whereas the passive tag size and ability to read near water or metal decrease with increase in frequency.
- Operating frequency, read performance, ruggedness, compliance with standards, operating conditions, and ability to upgrade are some main factors considered in selecting hardware components such as tags and readers.
- The selection of an RFID portal (where readers will be mounted) depends on the type of tracking (case level or item level) and the application. The goal is to maximize read performance.

Summary

You'll want to design your RFID system to meet application performance requirements. The task of designing your RFID system includes selecting the operating frequency; selecting hardware components such as readers, tags, and antennas; and selecting portals where the readers will be mounted to read the tags.

The three main factors that help you select the operating frequency for your RFID system are application type, required read range, and operating conditions. Some main factors considered in selecting readers and tags are operating frequency, read performance, ruggedness, compliance with standards, operating conditions, and ability to upgrade. The selection of an RFID portal (where readers will be mounted) depends on the type of tracking (case level or item level) and the application. The goal here is to maximize read performance. Some examples of pallet-level read portals are dock doors and forklifts, whereas an example of a portal for case-level reading is a conveyor. Some examples of item-level portals are smart shelves and POS systems.

After you select the operating frequency for your RFID system, you will need to identify the potential sources of interference around this frequency. In other words, your design is not final until you do a site analysis, which we'll discuss in the next chapter.

Exam's Eye View

Comprehend

- Frequency and wavelength are inversely proportional to each other; in other words, with an increase in frequency, the wavelength decreases.
- In general, with an increase in operating frequency, the read range increases.
- The read range of active tags is higher than that of passive tags.
- Do not unnecessarily use higher-frequency devices, and do not unnecessarily increase the maximum read distance, because higher frequencies and larger read distances are more vulnerable to interference with the surroundings, which will impede performance.
- Due to the higher wavelength, LF signals are not easily absorbed by the atmosphere and the material they move through. For this reason, RFID systems operating in the LF range work well around water and metal.
- The three main factors that help you select the operating frequency for your RFID system are application type, required read range, and operating conditions.

Look Out

- The only globally accepted radio frequency for RFID systems is 13.56 MHz, which falls in the HF band.
- Due to the short read range and less absorption, the LF systems are more robust to external influences.
- A circularly polarized reader antenna is preferred in applications in which the tag orientation is unknown or unpredictable, whereas linearly polarized antennas are useful for applications in which the tag orientation is fixed and known.

Memorize

- For applications that require reading the tags from a close proximity of a few centimeters to less than a meter (a few inches to less than 3 feet), use LF (125 or 134 KHz) or HF (13.56 MHz) RFID devices (tags and readers).
- Read range and read speed increase with an increase in frequency, whereas the passive tag size and ability to read near water or metal decrease with an increase in frequency.
- Read range depends on operating frequency, tag type (active or passive), communication technique (inductive coupling or backscattering), and radiated power.
- Most readers support a minimum of two and a maximum of four antenna ports.
- Two cable types commonly used as transmission lines in RFID systems are coaxial cable and shielded pair cable.
- Conveyors are used for case-level tracking, whereas dock doors and forklifts are good for pallet-level reading.

Key Terms

Antenna A structure or device used to receive or radiate electromagnetic waves.

Circularly polarized antenna An antenna that radiates circularly polarized waves, which are waves in which the electric field vector rotates in a circle as the waves travel.

Dipole antenna An antenna that consists of a straight electric conductor made of conducting metal such as copper, interrupted at the center, and therefore making two poles.

High frequency (HF) The frequency band of 3–30 MHz. RFID systems in this band operate at 13.56 MHz.

Horizontal polarization Linear polarization in which the wave travels horizontal to the surface of the Earth.

Isotropic antenna A hypothetical nondirectional antenna that radiates power uniformly in all directions. It is often used as a reference to calculate quantities such as effective radiated power (ERP).

Linearly polarized antenna An antenna that radiates linearly polarized waves, which are waves in which the electric field vector stays parallel to a line in space as the waves travel.

Low frequency (LF) The frequency band of 30–300 KHz. RFID systems in this band typically operate at 125 KHz or 134 KHz.

Microwave frequency A frequency band of 1 GHz–300 GHz. RFID systems in this band typically operate at 2.44 GHz or 5.80 GHz.

Omnidirectional antenna An omnidirectional antenna is a nondirectional antenna that radiates power uniformly in all directions. An ideally perfect omnidirectional antenna is also called an *isotropic antenna.*

Polarization The property of EM waves such as RF waves that determines the direction of the electric field in the plane perpendicular to the direction of wave propagation.

Reader The RFID component that communicates with the tag to receive information about the tagged item and sends this information to a host system. Readers are also called *interrogators*.

Read range The maximum distance from which a tag can be read.

Reading efficiency The ratio of the number of successful reads to the total number of read attempts.

Reading speed The number of tags a reader can read per unit of time.

RFID portal The area where RFID tags can be read or written to.

Ultrahigh frequency (UHF) A frequency band of 300 MHz–3GHz. RFID devices in this band operate at different frequencies in different regions of the world.

Vertical polarization Linear polarization in which the wave travels perpendicular to the surface of the Earth.

Self Test

A Quick Answer Key follows the Self Test questions. For complete answers and explanations to the Self Test questions in this chapter as well as the other chapters in this book, see **Appendix A**.

1. A perfectly omnidirectional antenna is called:
   A. An isotropic antenna
   B. A helical antenna
   C. A circularly polarized antenna
   D. A unidirectional antenna
2. The three-dimensional area that consists of RF waves being radiated from a reader antenna is called:
   A. The interrogation zone
   B. The footprint
   C. The radiation pattern
   D. Circular polarization
3. Which of the following correctly describes the reading efficiency or read robustness?
   A. Read range
   B. Ratio of number of successful reads to number of read attempts
   C. Number of read attempts
   D. The signal strength
4. What is the name of the protocol that the readers in an RFID system support so that they can be managed remotely?
   A. RemoteMan
   B. SMTP
   C. SNMP
   D. HTTP

5. You can use a coaxial cable as a transmission line in an RFID system for a maximum operating frequency of:

   A. 125 KHz

   B. 450 MHz

   C. 1 GHz

   D. 3 GHz

6. Match the items in the second column of the table to the items in the first column.

Mounting Equipment (Portal)	Function
A. Conveyor	1. Helps reduce out-of-stock situations
B. Dock door	2. Provides pallet-level tracking; requires operator intervention
C. Forklift	3. Provides pallet-level tracking and works in the presence of other electronic devices
D. Smart shelf	4. Provides case-level tracking and works better with multiple antennas

7. Which of the following is *not* an advantage of using a UHF RFID system?

   A. The signals will penetrate through materials such as water

   B. High data capacity

   C. Larger read range

   D. High data transfer rate

8. For which of the following applications will passive UFH tags not be suitable? (Choose one.)

   A. Case-level tracking at the airport

   B. Pallet-level tracking in a warehouse

   C. A real-time locating system (RTLS)

D. Product tracking in a supply chain application

9. Which will be the most suitable operating frequency for an application that requires a close read of items with high water content?

A. LF

B. HF

C. UHF

D. Microwave

Self Test Quick Answer Key

For complete answers and explanations to the Self Test questions in this chapter as well as the other chapters in this book, see Appendix A.

1. A
2. B
3. B
4. C
5. D
6. A-4
   B-3
   C-2
   D-1
7. A
8. C
9. A

Chapter 7

RFID+

Performing Site Analysis

Exam Objectives	What It Really Means
7.1 Given a scenario, demonstrate how to read blueprints (e.g., whole infrastructure)	Understand that blueprints are both input and output of the site analysis. You must understand what information you can find in a blueprint to start the site analysis, and what information you should enter into the blueprint after finishing your site analysis.
7.3 Given a scenario, analyze environmental conditions end-to-end	Understand what is included in the physical environmental analysis of the site.
7.2 Determine sources of interference	Know the different types and sources of EM noise and interference. You must know what the ambient EM noise (AEN) is and how to measure it. You must know the spectrum analyzer can be sued to measure AEN and interference. You should also know which variables are affected by the interferences and what are the possible solutions.

Introduction

The RFID system that you are designing will most probably be installed in an already existing infrastructure that contains other systems and devices. You need to determine how this RFID system will fit into that existing site infrastructure. For this reason, a site analysis is required before you finalize the RFID system design and before you install the RFID system. To aid you in this task, you can use a *blueprint* to visualize the site's physical infrastructure. You will need to analyze the site's physical infrastructure and RF environment to find appropriate locations for the interrogation zones or to mark the planned interrogation zones.

The main goal of the site analysis is to ensure that the interrogation zones will function properly, with maximum performance and without interrupting the existing services. Depending on the situation, you could have no freedom in choosing these zones, full freedom in choosing these zones, or somewhere in between these two extremes. Even if you have no freedom in choosing a zone, there could still be a degree of freedom to decide where exactly in the zone you will mount the reader antenna. So, regardless of how much freedom you have to determine the location of interrogation zones, a site analysis will be useful and required.

So, the core question in this chapter is: How can we perform a successful site analysis? In search of an answer, we will explore three avenues: performing physical environmental analysis, performing RF environmental analysis, and documenting and using our findings.

Planning the Site Analysis

Due to its very nature, an RFID system is almost always installed in an already existing infrastructure that contains other systems and devices. The purpose of the site analysis is to determine how your RFID system will fit into an already existing world at the site. To ensure that all aspects of the site analysis are addressed and to optimize your results, treat your site analysis like a project and therefore plan it. Planning a site analysis includes determining what this analysis will include and what will be its deliverables.

Plan the Steps Ahead

So, what is involved in performing a site analysis? This might slightly depend on the site and the requirements of the RFID system that will be installed. However, in general, your site analysis will include the following steps:

1. **Plan blueprints** Arrange the blueprints, which are basically the site diagrams that you will need to visualize the site infrastructure. You will also develop (or modify) blueprints at the end of your project to include your findings. So, blueprints are input to the site analysis project, and they are also the project deliverables.

2. **Inspect the site** Inspecting the site includes walking through the site and making observations for physical and electrical environmental analysis—for example, taking notes on any potential obstructions that can reflect or block the RF signal.
3. **Determine interrogation zones** This involves determining the spots where you can mount readers for reading tags. This step will help you avoid installing expensive readers or antennas at places where they will not be effective or where they will not be needed. If the location for an interrogation zone is already fixed, you will need to determine where exactly in that location you will install the reader antenna.
4. **Document your results** You will document the results of your site analysis in reports and in the form of blueprints. These are the deliverables of your site analysis project.

If you want to hit the ground running, you need to understand what blueprints are.

Understanding Blueprints

The first step of a site survey is to obtain the facility diagram, also called a *blueprint*. A blueprint, in general, is any plan that documents an architecture or an engineering design. A site blueprint helps you visualize the big picture of the site infrastructure. It will help you make a preliminary determination of where you can possibly set up the interrogation zones. Typically, standard symbols are used on the blueprint to represent various items. Figures 7.1 and 7.2 show very simple examples of a warehouse blueprint and some electrical and telecom symbols commonly used in blueprints.

The blueprint will give you the preliminary idea about the locations that you want to avoid while setting up interrogator zones, such as metallic material, and the things that you can use, such as breakout boxes, circuit breakers, and power drops. You also need to consider power availability, backup power source, availability of dedicated power circuits, and location of power outlets.

So, the blueprint shows you what physical infrastructure is in place. But it's only the starting point. Next, you need to visit the facility and analyze the physical environment of the site.

Figure 7.1 A Very Simple and Incomplete Illustration of a Warehouse Blueprint

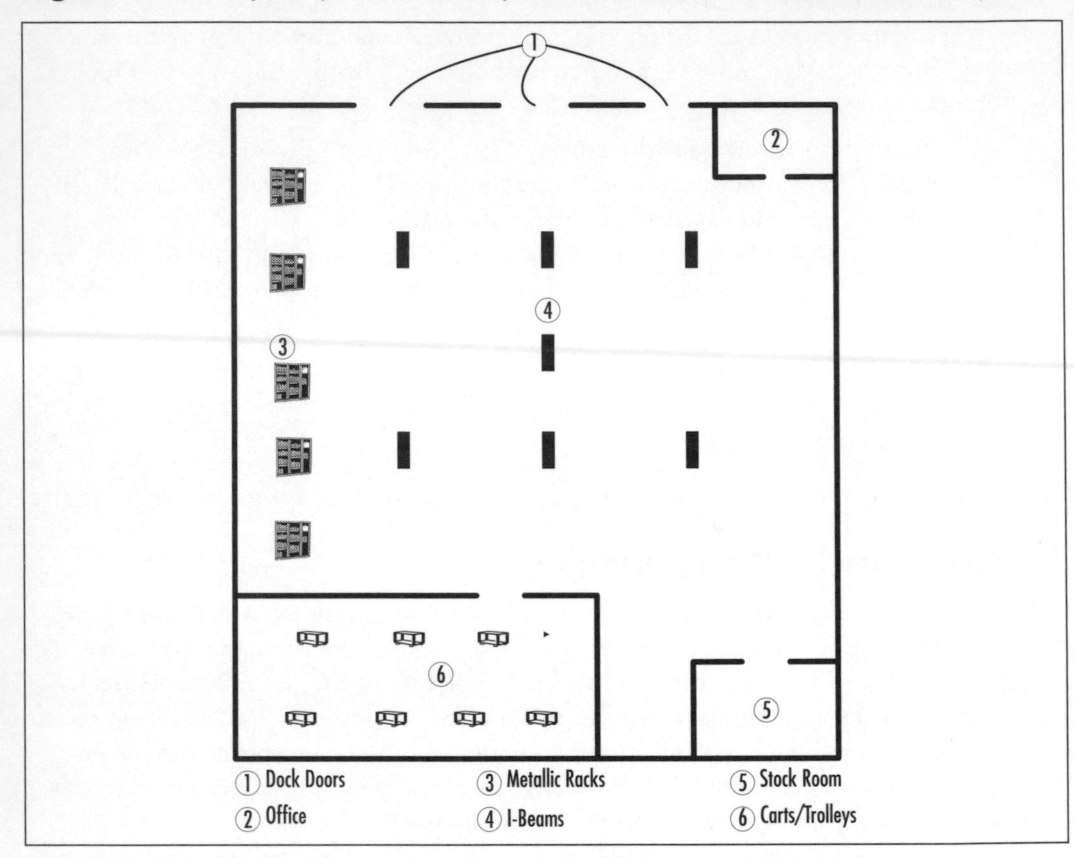

Figure 7.2 Examples of Electrical and Telecom Symbols Commonly Used in Blueprints

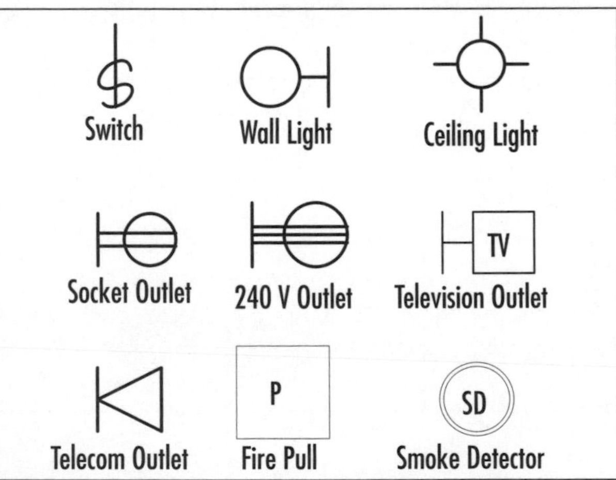

Performing a Physical Environmental Analysis

Mounting interrogators and thereby setting up interrogation zones is an important task in deploying an RFID system. The interrogator zones are set up in a physical space. Therefore, you first want to analyze the physical environment in which you are going to set up the interrogation zones. The physical environment that you are concerned with may consist of hindering characteristics or objects, such as extreme temperatures, moisture, high-power machinery, metallic equipment, and lighting fixtures. To perform the physical environmental analysis, you might need to consider various physical conditions described in the following sections.

Harsh Environmental Conditions

While setting up interrogation zones, you should consider harsh environmental conditions, including corrosive chemicals, extremely high or low temperature, humidity, and mechanical shocks, which can damage tags. Moisture and water contents can absorb RF signals emitted by readers and tags.

Physical Obstructions

A physical obstruction is an object that blocks the communication path between an interrogator and a reader. Examples of physical obstructions are beams, equipment, products, and electric motors. Depending on the nature of the physical obstruction, it can create adverse effects such as absorption, reflection, scattering, and interference.

Metallic Material

Metallic objects reflect RF signals, and that reflection can cause interference with the incident signal. You need to keep metallic objects such as casings outside the interrogation zones. The metallic parts of the RF system itself will also reflect RF energy. The solutions are to not choose the metallic part when you have an option and to create air gaps or use nonmetallic spacers to separate the metal. You might also need to adjust the distance between a reader and a tag to avoid the reflection effect.

TIP

All antennas and interrogators are securely mounted with the appropriate hardware. You must consider adverse effects such as reflection and interference while you're selecting the hardware to hold the equipment. For example, avoid using metal brackets.

Packaging

Given its nature, packaging can absorb, reflect, or scatter RF energy and thereby hamper the performance of an RFID system.

Cabling

You should make sure that the power and network cables will stay away from the interrogator zones, to avoid some adverse effects such as noise and interference.

Electrostatic Discharge

Electrostatic discharge, or ESD, is the instantaneous electric current created by the flow of electrons from a high-density (of electrons) surface to a low-density surface—for example, when the two surfaces rub against each other. ESD could gradually degrade (damage) a system component. In the environment of an RFID system, the common sources of ESD include belts, conveyors, rollers, paper handling, and striping labels from rolls. ESD can damage the transistors in a tag's IC, thereby causing the tag to malfunction. ESD can also damage the interrogator's IC, especially if the interrogator is not properly grounded.

EXERCISE 7.1

You are tagging the metallic by attaching tags to them. You know that the metallic surface will reflect the RF signal. What can you do to improve the performance?

Solution:

You can separate the metal from the tag by placing nonmetallic spacers that will create air gaps between the tags and the metallic surface. This can improve performance by increasing the read range.

You have visited the site and have recorded all the elements in the physical environment relevant to your system. This is important. But the real enemy is still lurking in the air: the RF waves waiting to attack your RFID system in the form of noise and interference.

Performing an RF Environmental Analysis

The RF environmental analysis includes identifying the sources of EM noise and interference that can hamper the performance of the proposed RFID system. The increasing use of EM waves to communicate has revolutionized communication capabilities and options to include cordless communication devices, satellite communication systems, and wireless networks. Looking at the other side of the coin, the increase in EM radiation in the environment can cause these waves to interfere with each other, which will disrupt operations and services. The purpose of the RF environmental analysis is to identify the potential sources of interference at and around the selected operating frequency of the proposed RFID system. If these sources stay unidentified and therefore the resulting problems unsolved, the interference will disrupt the communication between the readers and the tags, and it will consume lots of your time and resources to troubleshoot the problems of the running system.

The purpose of the RF environmental analysis is twofold:

- Eliminate any interference inside the interrogation zone.
- Ensure that the RFID system (tags and interrogators) will not interfere with the existing RF systems on the site.

TIP

If you are a beginner, consider accepting the help of an experienced RF engineer in performing the RF site survey. After going through trials and errors during multiple surveys, you will start feeling comfortable with doing it yourself.

Interference can have adverse effect on the following characteristics of an RFID system:

- Read speed
- Accuracy of communication
- Read range

You are looking for the EM systems on the site that can generate signals (or noise) in or around the same frequency range as that of the proposed RFID system and thereby cause signal interference. To deal with these electrical environmental conditions, you need to perform the following tasks:

- Identify the interference sources.
- Understand different interference types.
- Determine the ambient noise.
- Analyze these electrical environmental conditions.
- Design solutions for the discovered problems.

All these tasks are explained in the following sections. First, to identify the sources of this interference, you need to perform a site test before deploying the RFID system. This site test is called a *site assessment* or *site survey*.

Planning a Site Survey

Following are the apparatus that you will need to perform a successful site survey:

- A blueprint or a computer-aided drawing (CAD) drawing to visualize the site infrastructure
- An antenna that will cover 360 degrees of RF field, such as a half-wavelength dipole antenna
- A spectrum analyzer to measure noise and interfering signals
- A portable computer such as a laptop to record the collected data
- Two stands, such as tripod stands, to support the antennas
- A cart on which you can move your testing equipment around the site

Interference can come from a variety of sources. You should be aware of the types of interference that can come from the environment to impair the communication between readers and tags (or any RF communication, for that matter). Following are the common types of such interference:

- **Adjacent channel interference** This is the interference from a signal with a frequency close to the operating frequency of the RFID system. So, this is the interference between two frequency channels (bands).
- **Band congestion interference** This is the interference resulting from the overcrowding of a given frequency band—that is, too many devices operating within a shared frequency band.
- **Environmental interference** This is the interference from natural sources of EM radiation, such as lightning and solar radiation.
- **Jamming** This is the interference or the noise caused by an intentional emission of radiation by another device or system. This is done to limit the effectiveness of the other communications or detection equipment. For example, cellular phone jammers are used in locations where a phone call will be disruptive, such as in libraries and movie theaters.
- **Spurious emissions interference** Spurious emissions are the interfering radiation transmitted outside the operating frequency band in the form of narrowband signals or wideband noise. One example of spurious emissions is the emission of harmonics at multiples of fundamental frequency.

NOTE

The *harmonic* of a wave is a component frequency of the signal that is an integer multiple of the fundamental frequency.

To fully estimate the risk and impact of the RF environment, knowledge of the ambient noise is of particular importance.

Determining the Ambient EM Noise

The *ambient EM noise (AEN)* is the EM noise existing at a given location, such as a compartment, a room, or a particular outdoor location. AEN is generated by electrical devices such as infrared scanners, real-time location systems (RTLS), and alarm motion detectors. The AEN is measured in dB relative to some reference value—for example, the average power.

NOTE

The term *ambient* refers to the immediate surroundings of something. It comes from the French word *ambiant* and its roots go further back, to Latin.

The AEN can be measured by using a device called *spectrum analyzer*, which in general is used to examine the spectral composition of a an EM wave. Figure 7.3 shows an example of output from a spectrum analyzer: the Agilent 89600 Vector Signal Analyzer, a sophisticated device that displays the overall RF spectrum as well as the analyses of its various aspects. You can use a spectrum analyzer to perform the following tasks:

- Identify the source of RF interference.
- Measure the RF output from circuits, devices, and instruments.
- Measure the distortion, harmonic content, modulation quality, and noise.
- Display signal interference if it overlaps the intended signal.

Figure 7.3 An Example of a Spectrum Analyzer: The Agilent 89600 Vector Signal Analyzer *(Image Courtesy of Agilent Technologies, Inc.)*

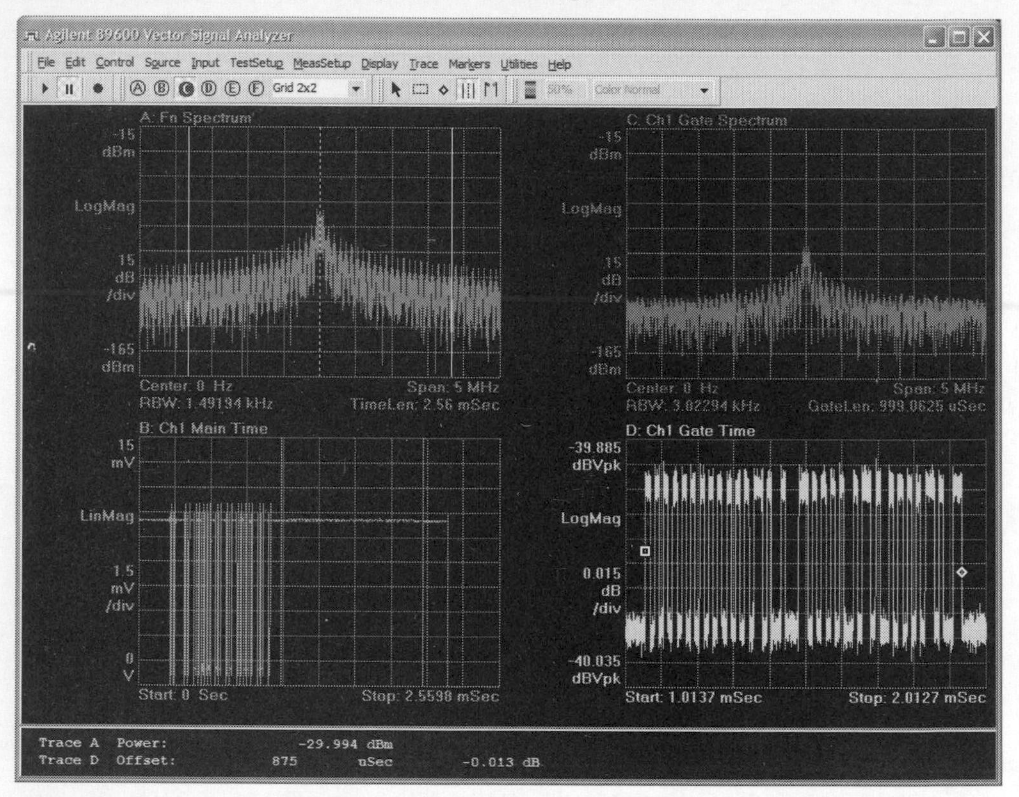

So, in general, you can use a spectrum analyzer to analyze the electrical (RF) environmental conditions and estimate their effect on the system variables, including the following:

- Antenna size
- Composition of the tagged object
- Operating frequency
- Power emitted by the interrogator

What should be the overall strategy to analyze the electrical environmental conditions, or the RF environment of the site?

Analyzing the Electrical Environmental Conditions

Performing an analysis of the electrical environmental conditions—that is, RF environmental analysis—includes the following:

- Identify the AEN.
- Measure the strength of interfering signals and noise. You need to identify these sources of interference and collect data with the help of a spectrum analyzer over a full operational business cycle, which typically is a minimum of 24 to 48 hours. This analysis, which is called a *full Faraday cycle analysis (FFCA)*, will tell you what frequencies will cause the most problems when you operate your RFID system in your facility.
- Map the interrogation zones to the site drawings, such as blueprints, and ensure that these zones are noise free.
- After you identify a device that might create EM noise, turn on this device and measure and record the results in the interrogation zone nearby.

NOTE

To capture the RF signal, you can place an antenna in the middle of the interrogation zone and connect it to the input of the spectrum analyzer.

So, you have performed a site survey to identify the sources of interference and AEN, and you have estimated the effect of interference on the system variables. What will you do next?

Protecting the RFID System from Interference and Noise

Once you know that your system will have interference (or noise) problems, you can start designing the solutions. To eliminate or minimize the effects of interference, you can consider implementing the following solutions:

- **Remove the source of interference** If it is possible to remove the source of interference, go ahead and do it. For example, it might be possible to ensure that a forklift or an electronic scale causing the interference does not operate in the interrogation zone.

- **Shield the source or RFID system** You can shield the RFID system or components to protect them from intentional or accidental interference, or you can shield the source of interference.
- **Use filters** You can use filters, which permit only selected frequencies to pass through a connected device by rejecting all other frequencies.
- **Avoid grounding loops** A grounding point is used to ensure the safety of the equipment and the operator. It also provides immunity to interference and noise. However, a grounding point can create a loop condition, which will cause energy transfer back to the connected devices and thereby interference and noise. This can be avoided by ensuring that a loop condition does not exist—for example, by ensuring that the conductor used for grounding is short enough.

Just like any other project, the site analysis is not complete without its deliverables. One of the deliverables of this project will be your own blueprint, or site diagram, containing the RF information useful for the installation of the RFID system.

Preparing Your Own Blueprints

You started from a blueprint, inspected the site to observe the physical and electrical environment, and at the end the whole survey will narrow down to the interrogation zones. In other words, after performing the physical and electrical site survey, you need to make some tests on the planned interrogation zones. You should record your results in your own site blueprint and in a report. These results will be used during installation of your RFID system.

Let the Experiment Begin

Consider a scenario in which you want to determine the coverage areas of antennas (interrogation zones) at various locations inside a warehouse. Typically, a warehouse will have dock doors, metal shelves, a stockroom with liquid and metallic material, trailers, and walls. All these obstacles affect the RF signal in different ways and different amounts, which are not known in advance. So, tests are necessary to determine the RF coverage pattern near these places.

Perform the following steps to measure the RF coverage areas for the reader antenna at different spots in the warehouse:

1. Choose a spot and set up an antenna for which the coverage area needs to be measured.
2. Use a spectrum analyzer to measure the strength of the signal emitted by the antenna. Take readings at several points in the same direction, starting from near the antenna and gradually moving away from it.

3. Make sure you record your readings. You will note that the signal strength decreases with an increase in distance from the antenna.
4. When you move far enough so that the signal strength is below a useful level, mark that point.
5. Repeat Steps 2 through 4 as many times as necessary to take readings in different directions of interest. These readings will mark the interrogation zone around the antenna.
6. Repeat Steps 1 through 5 as many times as necessary to mark the coverage areas (interrogation zones) at different spots.

You may be testing the spots (for coverage area) near the office areas, trolleys, and metal shelves; and you will observe the effects. Record your results into the site blueprint by showing the locations—for example, with the following coverage grading:

- Maximum coverage
- Minimum coverage
- Intermediate coverage

You will notice that the signal coverage area will decrease around spots near obstructions, such as metallic equipment.

Once you have determined the location of an interrogator zone, you can perform on it what is called *path loss contour mapping (PLCM)* by performing *path loss contour analysis (PLCA)*. PLCA is the process of determining how the field strength and shape of the RF coverage in an interrogation zone varies. In other words, PLCA data has the information about how the RF signals (waves) are degraded and distorted and how the wavefront (the shape of coverage) changes throughout the interrogation zone. PLCM is the process of preparing a blueprint that maps the PLCA data. The PCLA helps determining the following deployment variables:

- The location of the antenna in the interrogation zone
- Antenna alignment
- Setting of the emitted power

So, PCLM is used to fine-tune and configure a given interrogation zone for optimal performance, whereas FFCA is used to locate an interrogation zone.

CAUTION

Know the difference between full Faraday cycle analysis (FFCA) and path loss contour mapping (PLCM). FFCA is performed to identify the sources of interference, to determine the interrogation zone locations, whereas PLCM is per-

formed to get insight into the details of a given interrogation zone so that it can be fine-tuned and configured for optimal performance.

You will also make sure that any potential barriers that can affect RF signal propagation are also entered into the blueprint (or some other document). Following are some examples:

- Doors
- Electrical connections
- Metallic equipment
- Shelves
- Liquids and areas of high humidity
- Sources of interference and noise

After you have recorded your findings into the blueprint or elsewhere, you can use them to make installation decisions to facilitate error-free data transfer between tags and readers.

Using the Results of Your Experiment

This is how you can use the results of your experiment during the installation and deployment of the RFID system:

- Choose an antenna location that is free from obstacles.
- For a given area where tagged objects will be placed (or passing through), choose a spot for the antenna that maximizes the propagation pattern.
- Mount the antenna high enough to increase the horizontal coverage by RF signals.
- Use the results from PLCA to fine-tune the antenna.
- Choose a mounting spot for a reader that can accommodate cables for data transfer and electrical connections.
- Shield devices that generate RF, to prevent radiation leakage.
- Ensure that the users in the area know of the possible radiation hazard, and take steps to prevent excessive human exposure to the radiation by following the regulations.

The three most important takeaways from this chapter are the following:

- Physical environmental analysis is used to identify physical obstructions that can hamper the performance of your RFID system.
- RF environmental analysis is used to identify the sources of EM noise and interference that will hamper the performance of your RFID system and to measure the RF coverage in the planned interrogation zones.
- You should document your findings with adequate details and start determining the solutions to the discovered problems.

Summary

The purpose of site analysis is to determine how the proposed RFID system will fit into the existing site infrastructure. A blueprint, the site diagram, helps you visualize the site infrastructure. With this document as a starting point, the site analysis project has three stages: physical environmental analysis, RF environmental analysis, and documenting the results of your analysis. The physical environmental analysis includes recording harsh environmental conditions, physical obstructions, metallic materials, and other physical sources that may have adverse effects on the RF signal propagation. The RF environmental analysis includes identifying the sources of interference and noise. You also need to measure the interference and noise in the planned interrogation zones. The spectrum analyzer is your device to make these measurements.

You should take signal strength measurements in the planned interrogation zones to mark the coverage area. You must document your findings with adequate details. Some of these findings, such as the RF coverage information, can go into the blueprints. You will use these findings during installation, which is discussed in the next chapter.

Exam's Eye View

Comprehend

- ☑ Site analysis has two components: physical environmental analysis and RF environmental analysis.
- ☑ Interference can have adverse effects on the following characteristics of an RFID system: read speed, accuracy of communication, and read range.
- ☑ To capture the RF signal, an antenna can be placed in the middle of the interrogation zone and connected to the input of the spectrum analyzer.
- ☑ You use blueprints to visualize the site infrastructure and then to enter some results of the site analysis. Therefore, the blueprint can be used before during and after site analysis.
- ☑ FFCA is performed to identify the sources of interference, to determine the interrogation zone locations, whereas PLCM is performed to get an insight into the details of a given interrogation zone so that it can be fine-tuned and configured for optimal performance.

Look Out

- ☑ Collect the AEN data over a full Faraday cycle covering all normal business operations. Do not simply be satisfied with the AEN snapshot taken at a particular time.
- ☑ You must consider adverse effects such as reflection and interference while selecting the hardware to hold the equipment. For example, avoid using metal brackets.
- ☑ It might sound obvious, but document your findings with appropriate detail.
- ☑ While measuring the interference results in the interrogation zones, make sure you turn on the sources of interference.
- ☑ The site analysis, to some extent, can be performed at all stages: before, during, and after the RFID installation.

Memorize

- ☑ A blueprint is a site diagram used to visualize site infrastructure. A blueprint, in general, is any plan that documents an architecture or an engineering design.
- ☑ Full Faraday cycle analysis (FFCA) is a method to collect data regarding the EM waves in a site environment over a full business cycle, typically 24 to 48 hours, which covers all the normal operations that use the RF band in which you are collecting the data.
- ☑ Possible solutions for interference and noise include the following:
- ☑ Remove or shield the source of interference.
- ☑ Shield the RFID system.
- ☑ Use filters.
- ☑ Avoid grounding loops.

Key Terms

Adjacent channel interference The interference from a signal with a frequency close to the operating frequency of the RFID system.

Ambient EM noise (AEN) The EM noise existing at a given location, such as a compartment, a room, or a particular outdoor location.

Band congestion interference The interference resulting from overcrowding of frequency bands—that is, too many devices operating within a shared frequency band or closely spaced frequency bands.

Blueprint A site diagram used to visualize the site infrastructure. A blueprint, in general, is any plan that documents an architectural or an engineering design.

Electrostatic discharge (ESD) The instantaneous electric current created by the flow of electrons from a high-density (of electrons) surface to a low-density surface—for example, when the two surfaces rub against each other.

Full Faraday cycle analysis (FFCA) A process to collect data regarding the EM waves in a site environment over a full business cycle, which is typically 24 to 48 hours. A business cycle in this case is the time that includes all the normal operations involving the frequency band in which you are collecting the data.

Interference The interaction between two waves. The signal wave can interact with other waves that it meets on the way to its destination. A resultant wave is produced as a result of interference, and the receiver receives the resultant wave.

Environmental interference The interference from natural sources of EM radiation, such as lightning and solar radiation.

Jamming The interference or noise caused by an intentional emission of radiation by another device or system.

Noise An unwanted electrical wave (or energy) present in a circuit or in a signal.

Path loss contour analysis (PLCA) The process of determining how the field strength and shape of the RF coverage in an interrogation zone varies. The PLCA data has information about how the RF signals (waves) are degraded and distorted and how the wavefront (the shape of coverage) changes throughout the interrogation zone.

Path loss contour mapping (PLCM) The process of preparing a blueprint that maps the PLCA data.

Spectrum analyzer A device used to examine the spectral composition of an EM wave. You can use this tool to measure signal strength, interference, and AEN.

Spurious emissions The interfering radiation transmitted outside the operating frequency band in the form of narrowband signals or wideband noise.

Self Test

A Quick Answer Key follows the Self Test questions. For answers and explanations to the Self Test questions in this chapter as well as the other chapters in this book, see **Appendix A**.

1. Match the items in the second column of the table to the items in the first column.

Mounting Equipment (Portal)	Function
A. AEN	1. Helps visualize the infrastructure of the site
B. Site blueprint	2. Identifies the sources of interference by measuring the signal strength
C. Site survey	3. Ambient electromagnetic noise
D. Spectrum analyzer	4. Testing the site to identify sources of signal interference

2. Which of the following is used to mark the interrogation zone?
   A. Site map
   B. Site blueprint
   C. Spectrum analyzer
   D. Spectroscope
3. While planning for the site analysis, you need all of the following except:
   A. A cart
   B. A harmonic content analyzer
   C. Site blueprints
   D. A portable computer
4. You are panning for a site analysis and you have not yet walked through the site. You just want to visualize the site infrastructure. What tool will you use?
   A. Spectrum analyzer
   B. Site blueprint
   C. Site map
   D. Interrogation zone

5. The process of taking the RF data over a full business cycle during the site analysis is called:

   A. Full business cycle analysis

   B. The interrogation cycle

   C. Full Faraday cycle analysis

   D. The blueprint cycle

6. You have implemented an RFID system in a shipping area. Some of the items that you are going to tag have metallic surfaces. Which of the following solutions can you implement to optimize the read performance?

   A. Change the operating frequency of the tags

   B. Change the operating frequency of the readers

   C. Increase the power of the reader

   D. Use nonmetallic spacers to create air gaps between the tag and the metal

7. You are looking at the blueprint of a warehouse to visualize the site infrastructure before the site survey. Which of the following areas will have maximum signal coverage?

   A. Dock doors

   B. Metal equipment

   C. Cleaning room with liquids

   D. Trolleys

8. All of the following are purposes of site analysis except:

   A. Identifying sources of interference

   B. Marking interrogation zones by measuring signal strength

   C. Performing the cost analysis

   D. Finalizing system design

9. All of the following are possible solutions for protecting RFID systems from interference except:

   A. Removing the sources of interference

   B. Increasing the antenna size

   C. Using filters to permit only selected frequencies

   D. Avoiding grounding loops

Self Test Quick Answer Key

For answers and explanations to the Self Test questions in this chapter as well as the other chapters in this book, see Appendix A.

1. A-3
   B-1
   C-4
   D-2
2. C
3. B
4. B
5. C
6. D
7. A
8. C
9. B

Chapter 8

RFID+

Performing Installation

Exam Objectives	What It Really Means
6.1 Given a scenario, describe hardware installation using industry standard practices	Understand the standard process and practices to deploy an RFID system. You must understand that system design selection and site analysis is a prerequisite for a successful installation. Also understand the concepts of ESD, grounding, and grounding loop in context of hardware installation. You should know how to consider different system variables and power sources during installation. You must understand that safety is a requirement and not just an option.
6.2 Given a scenario, interpret a site diagram created by a RFID architect describing interrogation zone locations, cable drops, device mounting locations	You must be familiar with the issues involved with common installation scenarios such as conveyor portal, dock door portal, and forklift portal. You should also understand the importance of blueprints discussed in chapter 7.

Introduction

After selecting a system design, as discussed in Chapter 6, and performing a site analysis, as discussed in Chapter 7, you are now ready to install your RFID system. Think of installation as a process and not as an isolated task. The results from system design and site analysis are the input to the installation process. Installing hardware components and testing them are essential parts of deploying an RFID system. Before you actually install the hardware components, you should consider different installation scenarios and choose the one that best suits the given environment.

So, the core question in this chapter is: How can we successfully install an RFID system? In search of an answer, we will explore three avenues: planning installation, installing and testing hardware components, and considering various installation scenarios.

Preparing for Installation

Given the cost and the importance of an RFID system, its installation must be performed in a planned way to achieve the optimal results. An RFID installation is a process that begins when you start selecting the system design, which involves understanding various RFID solutions available in the context of application requirements. Equipped with that understanding, you perform the site analysis, which involves determining how the RFID system will fit into the existing site infrastructure. As Figure 8.1 depicts, the system design is not complete until the site analysis is performed. You use the information collected during system design and site analysis to install the system.

Figure 8.1 The Information Flow Among System Design Selection, Site Analysis, and Installation

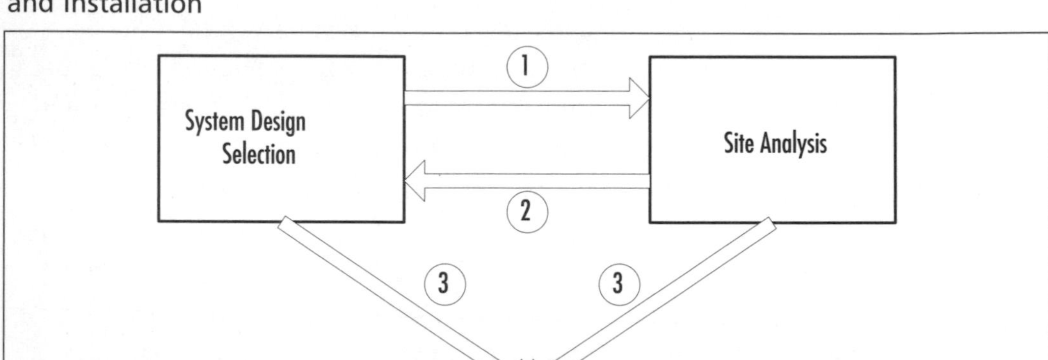

Preparing for installation includes once again considering all the system variables to put together an RFID solution, considering power sources for the system, and following the standard industry practices in installing the RFID system.

Putting Together an RFID Solution

Obtaining the components and installing an RFID system could be a very involved exercise. Although RFID is still an emerging and evolving technology, already several RFID solutions are available from multiple vendors. So what do you take and what do you leave? We discussed this issue in Chapter 7 to some extent. The bottom line is: It basically depends on the business requirements of the application for which you are deploying the RFID system. The requirements of the RFID system (the solution) that you install could also be influenced by the following factors:

- The site infrastructure
- The type of products that will be tracked by the RFID system
- The physical location where the RFID system is to be installed

You will make choices related to a number of important elements of the system; let's call them *system variables* because you have some freedom to choose them. After they are chosen, they will collectively define the system. These system variables that you must consider during installation are shown in Table 8.1.

Table 8.1 System Variables and Installation Considerations

System Variable	Considerations
Antenna	Number of antennas, antenna types, maximum power radiated, coverage area
Integration issues	Integrating the RFID system with existing systems such as applications and the network; some solutions might be easier to integrate than others
Maintenance	Some solutions might be easier to maintain than others
Operating frequency	Make sure the components are consistent with the selected operating frequency
Operating conditions	Make sure you obtain components that will withstand the operating conditions
Products to be tracked and identified (tagged)	The content, the packaging, and tag placement could affect whether the tags can be efficiently read
Regulations and standards	Safety regulations must be followed during installation; make sure the standards used by the components are consistent with each other

Continued

Table 8.1 continued System Variables and Installation Considerations

System Variable	Considerations
Readers and tags	The characteristics of readers and tags are consistent with each other; the readers will read the tags efficiently. Number of readers needed
Vendors	Some vendors might provide better documentation or customer service than others

Most of the variables listed in the table have already been discussed in previous chapters because they also need to be considered during the system design selection. Deploying an RFID solution includes system design, site analysis, and installation. It can also include hardware/software integration, configuration, and user training.

TIP

Various RFID vendors these days offer solutions for different scenarios, and some of them also offer installation services.

To operate, the system needs power. Where will the power come from?

Considering Power Sources

The readers and tags need power for their operation. Passive tags get their power from the RF signal emitted by the reader, whereas active tags have their own power source, typically a battery. Readers can use various power sources, including power supply units (PSUs), uninterrupted power supply (UPS), and power over Ethernet (POE).

Batteries

Active tags attached to items to be tracked and identified typically use batteries. Because a battery has a finite lifetime, it gives a tag a longer read range compared to passive tags, but they have a limited lifetime, typically up to 10 years. The application could determine what kind of battery should be used. For example, in toll applications, batteries should be able to withstand typical temperatures from –40° C to 80° C. In applications involving harsh environmental conditions such as tracking wildlife, the batteries used in the tags must be able to withstand even wider temperature ranges.

Power Supply Units

An *electrical power supply unit*, or PSU, is a device or system that supplies electrical energy to an output load or group of loads. The complete range of power supplies is very broad. You should be aware of the following two kinds:

- **Linear power supply** A *linear power supply unit* can be used as an AC-to-DC converter. A simple linear power supply unit powered with AC uses a transformer to convert the voltage—for example, from a wall outlet—to a lower DC voltage. Depending on the output load requirements, a linear regulator can be used to reduce the voltage to the desired output voltage. This power supply also provides other features such as current limiting.
- **Switched mode power supply** A *switching power supply unit* can be used as a DC-to-DC converter. In this case, the power supply is designed to accept a DC input from a limited range and to output a different DC voltage. This is especially useful in portable devices as well as for power distribution in large electronic equipment.

Uninterruptible Power Supplies

An *uninterruptible power supply*, or UPS, is a device that maintains a continuous supply of electric power to the connected equipment by supplying power from a battery when the utility power becomes unavailable. A UPS is inserted between the source of power and the output load that needs to be protected. When an abnormality such as a power failure occurs, the UPS automatically switches from the utility power to the battery power.

CAUTION

The switching power supply can generate noise, with some harmonics falling in the neighborhood of 125 KHz, which is the operating frequency for LF RFID systems. In that case, you might prefer to use a linear power supply.

Power Over Ethernet

The *power over Ethernet (POE)* technology system is any system that transmits electrical power along with data—for example, to remote devices over standard twisted-pair cable in an Ethernet network. This technology can be used for powering Ethernet hubs, IP telephones, Webcams, wireless access points, and other devices where it might not be convenient to supply power separately.

You should check out whether the company where you are going to install the RFID system already has implemented a power solution that you can use. For example, some organizations might already have implemented the power distribution solution based on access ports used in Ethernet cabling. This system can also be used to deliver data and power to RFID components such as interrogators.

You should understand the industry-standard process and practices for the installation and follow them.

The Standard Installation Process and Practices

Because RFID technology is still evolving, deployment of an RFID system can be a challenging task. You can minimize the risks involved by following the standard industry process for the installation based on best practices. It's important to consider installation as a process and not just as an isolated task. A process has an input, the actual tasks based on the input, and the output. In this case the input items are design selection and site analysis, the actual tasks are the installation tasks, and the output is the installed (deployed) system that will need to be managed. In the following sections we discuss all these elements.

Design Selection

The system design selection, discussed in Chapter 6, includes selecting operating frequency, hardware components, and types of RFID portal. This selection is largely driven by the application requirements and the environment in which the RFID system will operate. The environment is fully explored during site analysis, which is necessary to finalize the design.

Site Analysis

Site analysis is performed to determine how the proposed RFID system will fit into the existing site infrastructure. The process includes examining physical obstructions and electrical interference and noise. The goal is to mark the interrogation zones that can effectively coexist with the existing infrastructure. Equipped with the information from design selection and site analysis, you can start the installation tasks.

Installation Tasks

Installation tasks include installing hardware and possibly software components and testing the installed RFID system. While installing the system, you must consider and deal with three deployment issues:

- **Coexistence** The goal of a successful RFID implementation is that the RFID system effectively coexists with the existing site infrastructure. This means two things: The existing RF and other services are not disrupted by the RFID system, and the RFID system gets no or minimal interference or noise from the existing RF services. There is another dimension to coexistence: The RFID system components must work in harmony with one another. For example, you should consider the issues of interrogation zone overlapping.

- **Integration issues** Integration has two components: applications and network infrastructure.
  - **Integration with applications** The core RFID system (readers and interrogators) are usually integrated with the applications that analyze the data collected by the core system. For example, you might need to connect a reader to a host computer that might be supporting or connected to a database system. Internet communication could also be a part of the integrated system. Depending on the application, data aggregation and synchronization might be important issues in this case.
  - **Integration with an existing network** Multiple readers can be grouped together into a network to connect them to one or more host computers. A network (wired, wireless, or both) could already be in operation at the site, and you will need to integrate the RFID system to this network. The availability and reliability of the network connections will be important issues here.
- **System tuning** The readers and antennas are installed in different scenarios, such as conveyors, dock doors, and forklifts. In a given scenario, you should fine-tune the system after its installation to achieve the optimal results. The fine-tuning can include antenna positioning and alignment, adjusting antenna power emission within the legal limits, and so on.

System Management

During the installation, you must consider the system management-related issues; the RFID system will need to be managed after it is installed. Your choices during installation could affect system management after the installation is complete. Management includes the following components:

- **Cable management** You will need cables for connections, such as connecting readers to the power source, antennas to the readers, and readers to the network or host computer, and so on. These readers may be in the neighborhood of other equipments and operations, so it's important to consider the security and protection of the cables from possible damage and the safety of personnel from the cables.
- **Device management** There could be management issues related to the readers and the host computers attached to the readers. These issues can include network connection and data transmission.
- **Data management** You must anticipate what kind and volume of data is going to be collected by the RFID system. Ensure that the installed system is able to handle that data load.

The Tag Thing

You are installing the RFID system to read tags. Therefore, during the system installation, you must also consider what kind of tag the system will be reading and how the tags will be placed on the items.

You consider all these factors while planning your installation. After you've done the planning and made the required purchases, you are ready to start installing the hardware.

Installing Hardware

The first step in installing hardware is to compile at one place all the documentation, including information from the system design, information from the site analysis, and the manuals that came with the equipment that you purchased to install. You will be using this information throughout the installation process.

Installing hardware includes installing readers, installing antennas, and testing the interrogation zones. Here is the typical process for installing the core hardware for an RFID system:

1. Mount the reader.
2. Mount the antenna.
3. Install cables including connecting antenna to the reader.
4. Turn on the reader. Congratulations! You have an interrogation zone.
5. Test the interrogation zone.

CAUTION

During installation, keep your eyes open for any change that might have occurred in the site infrastructure after your site analysis. Such changes could require changes in your installation plan.

Now it's time to ask that important question: Where do I install the readers?

Installing Readers

As you know by now, a reader creates an interrogation zone in which it can read a tag. By the time of installation, you have already performed the site analysis to determine the interrogation zone. Recall that the location of an interrogation zone is determined by many factors, including the location of the tagged items to be read, sources of interference and noise, and sources of other adverse effects such as reflection and absorption. Once an interrogation zone has been marked, you have very little freedom in terms of where to mount the reader.

There are two main choices for mounting a reader. If the application requires it, a rack-based solution could be appropriate—for example, to withstand harsh environmental conditions. In this solution, a rack holds the antenna, reader, power supply such as UPS, and cables. The rack protects the system from harsh environmental conditions such as dust, humidity, and moisture.

The other solution is to mount the reader on some surface or edge, such as a wall or gantry placed around a conveyor. While using your freedom to choose the exact location to mount a reader, you must consider the following factors:

- **Breathing room** You will need clearance of a few inches around the reader for cables and to keep the air flowing so that the reader will stay cool.
- **Dense interrogator environment** By installation or configuration, or both, avoid the dense interrogator environment discussed in Chapter 4.
- **Environmental conditions** Avoid spots of harsh environmental conditions, such as extreme temperature, humidity, and moisture.
- **Safety** If there is lots of human movement in the area, choose a spot in which the reader will be safe from accidental physical damage.
- **Interference and noise** The sources of AEN and RF interference, discussed in the previous chapter, also play roles in determining the exact location of a reader. The idea here is obviously to avoid the effects of AEN and interference.

We will discuss some installation scenarios later in this chapter. The reader creates and processes) the signal, but the signal is transmitted and received by antennas connected to the reader.

Installing Antennas

Because antennas are communication elements, they are usually the most exposed components of an RFID system. Therefore, you must consider their protection from damage in your installation decisions. Obviously, an antenna will be mounted somewhere in or near the interrogation zone. RF path loss contour mapping (PLCM), discussed in the previous chapter, will help you determine the exact location of an antenna in a given interrogation zone and to configure and fine-tune the antenna. Depending on the situation, an antenna can be attached using drills and screws, or you might use a rack solution, as described previously.

Now you need to connect the antennas to their readers.

Installing Cables

An antenna is connected to a reader port via a cable. A reader typically has one, four, or eight antenna ports. Note that if the transmit and receive ports are separate, you will need two cables for each antenna, whereas if they are combined, only one cable per antenna will be needed.

You will also need cables for connecting other hardware components, such as power cables for the readers. Make sure you choose the right type of cable for each connection. Also consider possibly harsh conditions such as temperature or moisture (or mud) that the cables might need to put up with.

Use standard practices in installing cables, which including the following:

- Use the correct standard cable for a given connection: labeled by the manufacturer, authenticated by a quality organization for the purpose for which it's being used, or both.
- Keep the cables away from sources of EM waves such as motors and electronic devices.
- Use labels to identify the cables.
- Use protective caps to cover the exposed parts of the cables, such as cuts.
- The cable layout should be kept in an orderly fashion to avoid confusion and to have easy access for maintenance.

After you have set up an interrogation zone by installing a reader and its antenna and connecting them through a cable, you have an interrogation zone that you need to test.

Testing During Installation

Tests are the essential part of system deployment. There are four kinds of tests that you need to perform: interrogation zone tests, unit tests, application integration tests, and system tests.

Interrogation Zone Tests

After installing reader, antenna, and cables, make sure that the cables are connected correctly and that you are using the power supply appropriate for your purpose. Then turn the reader on.

Testing the interrogation zone involves three exercises:

1. Determine the boundaries of the coverage area (interrogation zone) by measuring the signal strength at various points around the antenna, as described in the previous chapter.
2. Verify the path loss contour mapping determined during site analysis, as described in the previous chapter. This involves making the field strength and signal strength measurements using a spectrum analyzer.

3. Use the path loss contour mapping to fine-tune and configure the RFID system. Test the system in a variety of configurations for optimal performance.

TIP

Take notes during installation and testing. These notes will help you troubleshoot problems during deployment and during the regular operation of the system.

Unit Tests

A unit test, in general, is a test procedure used to validate that a particular component of a system (software, hardware, or both) is functioning properly with the promised performance. For example, you should perform a unit test on a reader to verify that it meets performance specs, such as read range and multiple tag read rate, as specified by the vendor.

If the reader is connected to a host computer where the RFID data is to be used by an application, you might need to perform an application integration test.

Application Integration Tests

An application integration test is performed to verify that the RFID system works properly in collaboration with the application to which it is integrated. For example, the host computer might be sending requests to the RFID system to read the tags and send back the collected data, the data sent back to the host computer will be analyzed by an application, and so on. You need to test that the RFID system works harmoniously with the application.

System Tests

The term *system test* refers to testing the features of the collective system, including all the subsystems such as RFID subsystems and application subsystems. Depending on the application requirements, these tests can include different data load and capacity tests. Load tests verify that the system has the processing power to handle the amount of data it is expected to handle, whereas capacity tests verify the system's capability to produce the required output or results.

So, it's important to first test the units separately to verify their features and performance and then test the whole system put together. These are also called *preinstallation* and *post-installation tests.*

The safety factor must be included in the installation and test process.

Ensuring Safety

You must consider safety issues as a requirement rather than an option. When you are installing an RFID system, consider safety in the following three dimensions:

- Ensure that the RFID components are installed with proper care and protection to avoid possible damage to the components and to ensure the proper operation of the installed system.
- Ensure the safety of the personnel in the area.
- Ensure that the installed system conforms to safety regulations.

To cover all these safety dimensions, you must consider the following factors during installation.

Equipment Safety from the Environment

You should be mindful of the environment in which the equipment is being installed. Harsh environmental conditions such as extreme temperature, humidity, moisture, and condensation can have adverse effects on both the propagation of RF waves and the RFID equipment itself. These conditions can create adverse effects such as absorption of the RF waves and can damage the equipment. The two ways to protect against these conditions are by moving the interrogation zone away from them or using the enclosures for the RFID equipment when possible.

When you consider using enclosures, the name to remember is National Electrical Manufacturers Association (NEMA), an organization that provides standards and enclosures for electrical equipment. NEMA-rated enclosures come in various types, such as NEMA Type 4 and NEMA Type 12, each offering a set of protections. Some of the common protections offered by NEMA type enclosures include the following:

- Provide protection for personnel against accidental physical contact with the enclosed equipment.
- Provide protection for the enclosed equipment against material that may have adverse effects on the equipment, such as dripping water, dust, rain, sleet, snow, splashing water, hose-directed water?you've got the idea.
- Provide protection for the enclosed equipment against external formation of ice on the enclosure.
- Provide protection against corrosion, which is the deterioration of essential properties of a material object due to reactions with its environment. A common example of corrosion is rust.

TIP

All NEMA enclosure types are described in NEMA Standards Publication 250-1997: *Enclosures for Electrical Equipment (1000 Volts Maximum).*

Another source of possible damage that you need to protect against is electrostatic discharge.

Electrostatic Discharge

Electrostatic discharge, or ESD, is the instantaneous electric current created by the flow of electrons from a high-density (of electrons) surface to a low-density surface—for example, when the two surfaces rub against each other. It could gradually degrade and damage a system component. In the environment of an RFID system, common sources of ESD include belts, conveyors, rollers, paper handling, and striping labels from rolls.

Some ESD protection techniques are listed in the following:

- **Wrist straps** A wrist strap can be used as the primary method to ground personnel when they're close to an unprotected ESD item. The wrist strap must be worn in direct contact with the wearer's bare skin. Ensure that the ground connection end of the strap is securely connected to the ESD ground.
- **Groundable footwear** Groundable footwear can be used as an alternative to the wrist strap, especially in a situation in which wrist straps are not appropriate or are unsafe to use.
- **Conductive mats** Conductive mats can be used to ground personnel and furniture.
- **Plastic handles** While near ESD-sensitive items, use plastic-handled hand tools such as pliers, screwdrivers, and wire strippers.
- **Relative humidity** The relative humidity in the area where ESD protection is required must be above a minimum threshold value, say 30 percent. It must also be below an appropriate maximum limit because excessive humidity can cause problems such as corrosion, high-voltage leakage paths, and moisture contamination within the equipment.
- **Air ionization** This technique is used to neutralize charges on ungrounded conductors and insulators.
- **ESD-protective packaging and storage** The ESD sensitive items must be contained within approved ESD protective containers for movement in and between ESD-protected areas. When stored, ESD-sensitive items should be contained within a static-shielding container. Direct contact of unprotected

ESD items with metal shelves or cabinets must be avoided. Once the ESD-sensitive items stored in the cabinet are safely enclosed within ESD shielding, it is not necessary for the metal storage cabinets to be grounded.

CAUTION

Ordinary adhesive tapes such as a duct or masking tape can be highly chargeable. That means you should only use the approved tape near or in direct contact with an ESD-sensitive item. Do not use those tacky mats within a few meters (say, 3) of ESD-sensitive items. Also, unprotected ESD-sensitive items must not be passed from one individual to another unless both individuals are properly grounded.

ESD can damage the transistors in a tag's IC, thereby causing the tag to malfunction. ESD can also damage the interrogator's IC, especially if the interrogator is not properly grounded.

Grounding

Grounding refers to making electrical connection to the earth. The part directly in contact with the earth is called the *earth electrode*, and it can be as simple as a metal rod or a wire. Grounding provides a reference voltage level (zero) to electrical equipment and a sink to absorb electric charge under fault conditions. Grounding provides the following safety:

- It can dissipate electrostatic buildup.
- It is primarily used to prevent electric shock or fires caused by a voltage difference between the earth and a conducting material.
- It is also often used as a protection against lightning strikes because it will harmlessly conduct the resulting excessive charge to the earth rather than starting fires and damaging equipment.
- It is also used to control electrical noise and interference in electrical items such as computers, readers, and communication circuits.

CAUTION

An *electrical ground* must have enough charge-carrying (current) capability to serve as an effective zero-voltage reference level.

While dealing with grounding, you must be careful to avoid the possibilities of ground loops.

Ground Loops

Ground loops in an electrical system are unwanted currents that flow in a conductor connecting two points that are supposed to be at the same potential, e.g. ground (or zero) potential, but are actually at different potentials. Different points on Earth may have different electrical potential—the difference could be as big as hundreds of volts—for example, due to the influence of solar wind. Therefore, by grounding two components (say, readers) of a system to different points on Earth, you can create a ground loop. Such a condition could be very unsafe for a personnel operating or servicing the system.

A loop condition created by a ground point will cause energy transfer back to the connected devices and thereby will generate interference and noise. This can be avoided by ensuring that a loop condition does not exist—for example, by ensuring that the conductor used for grounding is short enough and that all the components (readers, in our case) are connected to the same grounding system.

Safety Regulations

During installation, you must recognize and conform to safety regulations regarding human exposure to the radiation emitted by the RFID system. Safety regulations and guidelines for human exposure to RF fields are necessary because if the RF energy absorption exceeds a threshold value, adverse biological effects could occur. This issue was discussed in detail in Chapter 5.

So, during the deployment of an RFID system, you must consider both kinds of safety: protection of the equipment from the environment and protection of the environment (including personnel) from the equipment. Toward this end, deploying an RFID system involves creating RFID portals.

Working with Various Installation Scenarios

When you are installing an RFID system, you are basically deploying what are called *RFID portals*. An RFID portal is an area in which tags can be read or written to. Note the difference between the two interrelated terms: portal and interrogation zone. When you mount a reader, you have portal. When you turn the reader on, you have an interrogation zone on the portal. From an installation viewpoint, the readers are divided into three categories:

- Fixed-mount interrogators
- Vehicle-mount interrogators
- Handheld interrogators

These categories were discussed in Chapter 4. Vehicle-mount and handheld interrogators are also collectively called *mobile interrogators*. So, most installation scenarios can be grouped into two categories: mobile reader installation and fixed-mount reader installation. While installing readers, always remember that you are installing an RFID portal, which is an area in which the tags cab be read or written to. Corresponding to fixed-mount and mobile readers, there are stationary ports and mobile ports.

Setting Up Stationary Portals

A stationary portal is set up by installing a fixed-mount reader. In fixed-mount reader scenarios, the tags go to the reader to be read. In other words, the reader waits for the tags to pass through its interrogation zone, and when they do, it reads them.

Setting Up a Conveyor Portal

Conveyors are used for case-level tracking—for example, in airports. This scenario has the following elements already fixed for you:

- The tags are moving at a certain speed.
- The tags are not oriented in a certain direction.

This situation requires that you use multiple readers for optimal results. The reader antennas are often mounted on gantries placed around the conveyor, as shown in Figure 8.2. The reader antennas on each side of the gantry will cover four faces of the container: up, down, and two side faces.

Figure 8.2 An Example of a Conveyor Portal with Four Antennas

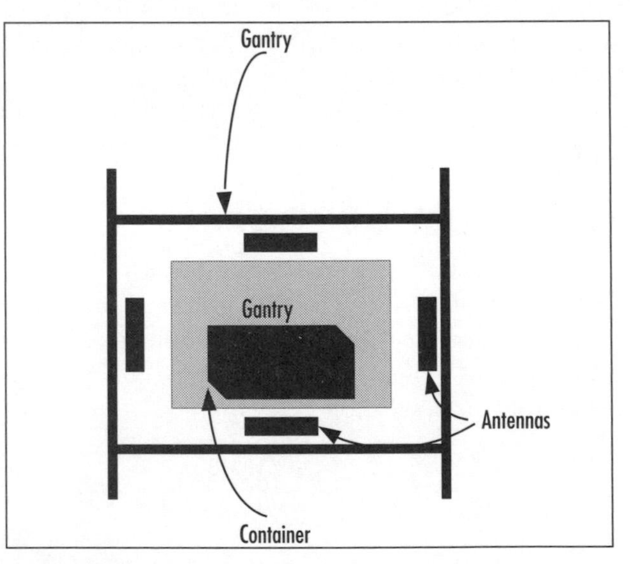

Other things that you should consider while setting up a conveyor portal include the following:

- To avoid signal reflection by metal and other adverse effects, keep the antennas comfortably away from the conveyor edges; about 45 cm (1.5 feet) away is a good rule of thumb.
- One of the four antennas will be positioned under the conveyor. You might find antennas designed for optimal operations under the conveyor.
- While configuring the reader for read-related parameters, keep in mind the speed of the conveyor (that is, as much time a tag has in the interrogation zone), which will also affect the number of attempts a reader can make to read a tag. A reader must be able to make multiple attempts to read a tag.
- The polarization and alignment of reader antennas should be compatible with random orientation of tags.
- While configuring the power emitted by the reader antenna, keep in mind the following two opposing factors:
  - There must be enough power to offer a large enough read range and interrogation zone.
  - The power must be kept within the safety regulations set by local, regional, national, and international regulatory bodies. This is important for protecting personnel in the area against radiation.

TIP

Because the conveyor is moving, a tag has only a limited amount of time in the interrogation zone. So, avoid writing to the tag on the conveyor, because it will consume some time. Also, the write requirements for power and distance are different from the read requirements.

So, conveyors pose a speed challenge to RFID systems. The higher the conveyor, the higher must be the reader's read speed. Readers have different read speeds, which can vary from about 25 tags/sec to well over 1000 tags/sec. However, you can find readers that can handle a conveyor speed of up to 600 feet/sec.

Exercise 8.1

You have adjusted the power of reader antennas set on a conveyor to a value that is within the limits of the safety regulations. However, it is higher than you need to cover the conveyor width. What adverse effect can it have on the tag reads?

Solution:

Due to the higher power, the read range is larger than required. The reader could start reading tags in the vicinity of the conveyor—for example, from a forklift truck passing by. Such a read is called a *stray tag read*.

Conveyors are good for case-level reading, whereas dock doors are suitable for pallet-level reading.

Setting Up a Dock Door Portal

Dock door portals are used for pallet-level tracking. They read the tags on items passing through a gateway such as a door. This scenario has the following elements already fixed for you:

- The region outside but near the door can be used for temporary staging of items for shipment. Those items must not be read by the readers in the door portal.
- The tag items will pass through the door—that is, they will be in the interrogation zone for only a short period.
- The way the tagged items are transported through the dock door, such as on a forklift, will determine the height at which an interrogation zone should be set up. The door size will also affect the determination of read range and interrogation zone.
- The dock door areas are usually busy with activities.

These situations require that you consider the following points while installing a dock door portal:

- Keep the cables out of harm's way.
- Make sure that the antennas are protected against accidental damage. You might consider a rack solution for antennas on two sides of the door.
- To offer the required read range, a UHF system is usually appropriate for this scenario.

- The antenna power should be high enough to offer the required read range but should be within the limits set by safety regulations. Also, too high a power can create stray read problems and interrogation zone overlaps.
- The number of antennas should be large enough to cover the width and the expected height at which the tagged items will be passing. Too many antennas could create the dense interrogator environment problem discussed in Chapter 4.
- Depending on the situation, there are several possible antenna placement solutions:
  - Two antennas on each side of the door: four antennas in total. A rack solution can be used if required for protection. An example of this scenario is shown in Figure 8.3.
  - Four antennas on each side of the door: eight antennas in total. A rack solution can be used if required for protection.
  - On antenna mounted overhead. This solution provides protection for the antenna.
  - Two antennas mounted overhead, looking into the door interior from the two corners of the door. This solution provides protection for the antennas.
- While positioning antennas, to avoid interference, make sure the antennas from two sides of the door are not directly pointing at each other.

Figure 8.3 An Example of a Dock Door Portal with Four Antennas

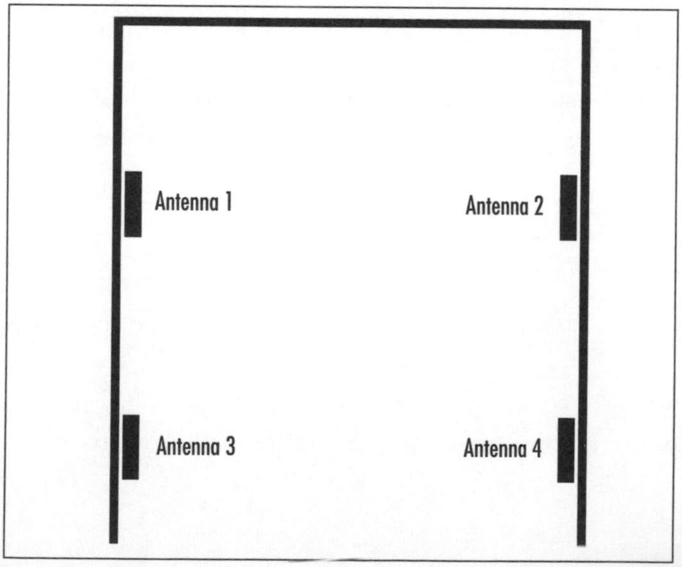

EXERCISE 8.2

You have set up RFID portals on two dock doors adjacent to each other. What are the two main problems that can happen in this scenario, and what are their possible solutions?

Problems: First, because two portals are just next to each other, the reader on one door could read tags passing through the other door—the stray tag read problem. Second, the interrogation zones of readers from two doors could overlap each other, which could create effects such as interference.

Possible solutions:

1. Configure and fine-tune the readers to remove these overlap problems.
2. Turn on only one portal at a given time. This can partially be achieved by a sensor-based portal: The portal readers are turned on only when a sensor on the door indicates something is coming and needs to be read.
3. Move the doors away from each other.

So, the main challenges posed by the dock door portal scenario are the following:

- Multiple dock doors adjacent to each other can create interrogation zone overlap problems, causing signal interference and reading tags from the other door.
- Because there could be items staged for shipping near the dock door, a reader on a dock door, if not properly configured and tuned, might read tags from the region outside the door—stray tag reads.
- The movement around a dock door poses the risk of damage to the RFID components. Therefore, the cables and the antennas must be protected against incidental damage.

Setting Up a Shelf Portal

This scenario has the following elements already fixed for you:

- The items on the shelf could have liquid content, and there might be metal in the vicinity.
- The number of items on the shelf keeps changing—that is, customers take the items away and more items are put into the shelf at a later time.

- The items will most probably be read from rather a close distance—say, less than 6 inches.

The most suitable RF to meet these requirements is HF—that is, 13.56 MHz. This frequency will give enough read range and will be relatively less affected by metals and liquid content—and therefore less vulnerable to effects such as absorption. Other things that you should consider as you install shelf portals include the following:

- You can create a multishelved portal with one antenna per shelf using a multi-antenna reader. In this case, you can configure the reader to use the antennas for reads in some sequential order.
- If the change in the number of items on the shelf needs to be detected, the readers should be configured to either continuously keep reading the tags or have read cycles at preset intervals. You need to consider the reading requirement while you're making the buying decision, because different readers from different vendors could offer different features and options.
- The reader antennas should be properly placed, configured, and oriented to avoid interference between neighbor antennas.

TIP

Unlike the dock door and conveyor portals, you have more control over the tags in a shelf portal. For optimal reads, make sure that the tags are oriented at right angles to the RF field coming from the reader antenna. If they are parallel to this field, no signal will be received, and hence no communication will occur between the reader and the tag.

So, the main issues in setting up a shelf portal are:

- Ensure proper orientation of the tags in the shelf so that they can be optimally read.
- Select the proper time interval between consecutive read cycles so that the change in items in the shelf can be optimally recorded.

Depending on the application requirements, you may also use the mobile reader to read tagged items in the shelves.

Setting Up Mobile Portals

A mobile portal is set up by installing a mobile reader. In the mobile reader scenario, the reader goes to the tag to read it.

Handheld Interrogator Portals

A handheld interrogator contains the whole reading system in one device and offers the maximum flexibility. You can take the interrogator to the tagged item and collect the information. Because handheld interrogators are usually used for close reads, interference from other RF devices and other adverse effects such as reflection and scattering due to neighbor objects are usually not issues. However, you still need to configure the reader according to the required read range.

Mobile-Mount Portals

A mobile mount portal is set up by mounting a reader on a vehicle such as a forklift. For example, you can collect information from tagged items on pallets in a storage room by driving an RFID-enabled forklift down the aisle. In a forklift scenario, the following parameters apply to the environment:

- Avoid harsh environmental conditions?for example, cold temperature in a storage room.
- Eliminate unfriendly material such as metals that will reflect the signal.
- You will need a wireless connection for the RFID system to send data to a central place such as a data warehouse.
- The reader will be in motion.
- You will need a power supply in the vehicle.

This scenario requires you to consider the following:

- You can mount the reader on either the interior or the exterior of the vehicle.
- In case you decide to mount the reader on the exterior of the vehicle, consider the adequate clearance from external objects such as doorways.
- The reader must be protected against mechanical shocks and vibrations.
- The mobile RFID system will most probably be using the wireless network for transferring data to a central location. Therefore, the possibility of interference with other EM devices should be considered.

So, the main challenges of a forklift-based mobile RFID portal are wireless network and harsh environmental conditions, including extreme temperatures, metallic material, mechanical shocks and vibrations, and so on.

Exercise 8.3

You are going to install RFID portals in a warehouse that has 75 dock doors and 10 forklifts. You have a choice to implement dock door stationary portals or forklift mobile portals. Which solution will be more cost effective, and why?

Solution: The forklift-based mobile portal solution will be more effective because you will need much fewer readers than you would for a dock door portal solution. For a dock door portal solution, you would need at least 75 readers for 75 dock doors. For 10 forklifts, you can manage with as few as 10 readers.

The three most important takeaways from this chapter are the following:

- Consider implementation as a process that has input actual implementation tasks and output. The input items are information from design selection and site analysis, the actual tasks are the installation tasks, and the output is the deployed system that will need to be managed.
- Hardware installation must include testing and conformance to safety regulations.
- A given installation scenario will have its own installation issues, and the emphasis on some installation issues might vary from one installation scenario to another.

Summary

Deployment of an RFID system is a process that takes information from system design and site analysis as an input, involves performing actual installation tasks, and considers the installed system as an output. For optimal results, the installation must be planned, which includes considering power sources and system variables such as operating frequency, operating environment, number of antennas, and so on. Hardware components such as readers and antennas must be tested as units for their features and performance before installing them into a system, where they will be tested again as part of the system test. While making installation decisions and during the actual installation, you must conform to the safety requirements that involve both protecting the equipment from the environment as well as protecting the environment (including personnel) from the equipment. You must consider all the possible installation scenarios available to you and choose the one that best suits the situation. Each installation scenario may have its own installation issues.

So, now you have implemented a basic RFID system. But your site might need additional devices such as an RFID printer to support the RFID system. We explore this topic in the next chapter.

Exam's Eye View

Comprehend

- ☑ Planning for installing an RFID system includes considering power sources and system variables such as operating frequency, operating conditions, number of antennas, and regulations and standards.
- ☑ Conveyors pose a speed challenge to the RFID systems. However, you can find readers that can meet speed requirements of up to 600 feet/sec.
- ☑ The main challenges that dock doors pose are the following:
- ☑ Readable items staged near the dock doors might be read by a reader on the dock door—a stray tag read.
- ☑ The interrogation zones of readers on the adjacent doors could overlap.
- ☑ The movement around the dock door poses a risk of damage to the RFID components.
- ☑ The main issues in setting up a shelf portal are:
- ☑ Ensuring proper orientation of the tags in the shelf so that they can be optimally read

- ☑ Selecting the proper time interval between consecutive read cycles so that the change in items in the shelf can be optimally recorded
- ☑ The main challenges in a forklift-based RFID portal are the wireless network and the harsh environmental conditions.

Look Out

- ☑ Increased antenna power could create the stray tag read problem.
- ☑ Too many reader antennas and increased power can create dense reader environment problems, such as interrogation zone overlap and signal interference.
- ☑ You must perform system design selection and site analysis before starting to install an RFID system.
- ☑ It's important to first test the units separately to verify their features and performance and then test the whole system put together.

Memorize

- ☑ HF (13.56 MHz) works best for a typical shelf portal, whereas UHF is more appropriate for a dock portal.
- ☑ The National Electrical Manufacturers Association (NEMA) provides standards for electrical equipment and enclosures for electrical equipment.
- ☑ Ground loops in an electrical system are unwanted currents that flow in a conductor connecting two points that are supposed to be at the same potential, e.g. ground (or zero) potential, but are actually at different potentials.
- ☑ You can find readers that can handle conveyor speeds of up to 600 feet/sec.

Key Terms

Corrosion The deterioration of essential properties of a material object due to reactions with its environment. A common example of corrosion is rust.

Electrostatic discharge (ESD) The instantaneous electric current created by the flow of electrons from a high-density (of electrons) surface to a low-density surface—for example, when the two surfaces rub against each other.

Ground Electrical connection to Earth. The part directly in contact with the Earth is called the Earth electrode, and it can be as simple as a metal rod or a wire. For example, multiple ground connection paths between two components of an electrical system will create a ground loop.

Ground loop An unwanted current that flows in a conductor connecting two points that are supposed to be at the same potential, e.g. ground (or zero) potential, but are actually at different potentials.

National Electrical Manufacturers Association (NEMA) An organization formed in 1926 from the merger of the Electric Power Club and the Associated Manufacturers of Electrical Supplies. NEMA provides a forum for the standardization of electrical equipment. It also helps the electrical industry by functioning as a central confidential agency for gathering, compiling, and analyzing market statistics and economic data.

Power over Ethernet (POE) A system that transmits electrical power along with data. For example, to remote devices over the standard twisted-pair cable in an Ethernet network.

Power supply unit (PSU) A device that supplies electrical energy to an output load or group of loads.

Reader speed The speed with which a reader collects information. It can vary from about 25 tags/sec to over 1000 tags/sec.

Relative humidity A quantity used to describe the amount of water vapor that exists in a gaseous mixture of air and water.

Stray tag read A read of a tag that is not supposed to be read by a reader. For example, due to high power, a reader can read tags outside its planned interrogation zone.

Uninterruptible power supply (UPS) A device that maintains a continuous supply of electric power to the connected equipment by supplying power from a battery when the utility power becomes unavailable.

Self Test

A Quick Answer Key follows the Self Test questions. For complete answers and explanations to the Self Test questions in this chapter as well as the other chapters in this book, see **Appendix A**.

1. Which RFID portal solution is the most flexible in terms of reading the tags from different distances and angles?

   A. A handheld interrogator

   B. A vehicle-mount interrogator

C. A dock door portal

D. A remote portal

2. Match the items in the second column of the table to the items in the first column.

RFID Portal	Most Important Function or Consideration
A. Conveyor	1. Stray tag read
B. Dock door	2. Mechanical shocks and vibrations
C. Forklift	3. Continuous read cycles
D. Shelf	4. Tag speed

3. Which organization provides standards for electrical equipment and enclosures for electrical equipment?

A. NEMA

B. EPCglobal

C. FCC

D. SAC

4. Which device do you use to verify the RF path loss contour?

A. An RFID encoder

B. An oscilloscope

C. A spectrum analyzer

D. A contour mapper

5. Which of the following power supplies maintains a continuous supply of electric power to the connected equipment by supplying power from a battery when the utility power becomes unavailable?

A. UPS

B. Linear power supply

C. Switched mode power supply

D. POE

6. Which of the following is not the way to protect RFID devices from ESD?

   A. Ground loops

   B. Dissipating electrostatic buildup through proper grounding

   C. Using ESD protective packaging

   D. Eliminating the generation of electrostatic discharge

7. A battery can be used as a power source in all of the following devices except:

   A. Real-time location systems (RTLS)

   B. A handheld reader

   C. A UHF passive tag

   D. Class 3 tags

8. RFID equipment tests:

   A. Should be performed before installation

   B. Should be performed after installation

   C. Are optional

   D. Should be performed before and after installation

9. You are installing an RFID portal on a door through which pallets will be passing. Which of the following is not a necessary step to take in applying safety?

   A. Ensure that the RFID components are installed with proper care and protection to avoid possible damage to the components and to ensure the proper operation of the installed system.

   B. If the reader antenna is emitting too much power, tell any security guard sitting near the door to change his location.

   C. Ensure the safety of the personnel in the area.

   D. Ensure that the installed system conforms to safety regulations.

Self Test Quick Answer Key

For complete answers and explanations to the Self Test questions in this chapter as well as the other chapters in this book, see the Appendix A.

1. C
2. D
3. B
4. D
5. C
6. A and B
7. B
8. D
9. B

Chapter 9

RFID+

Working with RFID Peripherals

Exam Objective	What It Really Means
9.1 Describe installation and configuration of RFID printer (may use scenarios)	Understand smart labels and smart label printer. You must understand how a smart label and smart label printer integrates the traditional barcode technology with RFID technology. You must know the basic installation procedure and the typical properties of the printer that you can configure. You should also understand how the printer handles the errors and how you can troubleshoot the printer encoder.
9.2 Describe ancillary devices/concepts	Understand how automatic label applicators work. You must understand how the automatic label applicators use different label placing techniques: tamp down, blow on, and wipe-on. You must also know how the feedback devices such as photo eye, horn, light tree, and motion sensor help build robust, effective, and automatic RFID systems and networks.

Introduction

RFID begins where barcode technology ends. A so-called *smart label* combines a\ barcode with an RFID tag. Smart labels have the intelligence and functionality of a tag and the printing convenience of a barcode label. Therefore, by using smart labels, corporations can leverage their existing labeling infrastructure to incorporate RFID.

Smart labels are created by *RFID printers*, which are peripheral devices, and they are applied to the items to be tracked by using another peripheral device called an *automatic label applicator*. The third kind of peripheral or ancillary device used in RFID networks is called a *feedback system*, which helps build effective, robust, and automatic RFID networks.

So, the core issue in this chapter is the role of peripheral devices in RFID. To understand this issue, we will explore three avenues: RFID printers, RFID label applicators, and RFID feedback devices.

Smart Labels: Where RFID Meets Barcode

Upon the arrival of RFID, the industry was using the barcode technology to identify items. A *barcode* (also written as *bar code*) is a machine-readable representation of information printed on a surface in a visual format. Almost everything that you buy from retailers these days has a barcode printed on it, which helps manufacturers and retailers keep track of inventory. Barcodes can be read by optical scanners, also called *barcode readers*. These codes serve as product "fingerprints" made of machine-readable, parallel bars that store binary code, as shown in Figure 9.1.

Figure 9.1 An Example of a Barcode

NOTE

Barcode labels are also called *Universal Product Code (UPC) labels* or *UPC barcodes*. UPC is the encoding scheme, or data structure, used to write barcodes, widely used in the United States and Canada on items in retail stores.

RFID tags offer what barcodes offer, plus lot more. For example, barcodes, once created, can only be read, whereas you can modify the information on the writable tags. Furthermore, the tags can store a lot more information, and they can become part of the global network. However, the procedures and infrastructure for barcodes (also called *labels*) is already in place. The incorporation of RFID tags into the existing labeling (barcode) technology gave rise to the smart label, which is basically a barcode label that contains an RFID tag embedded in it. It's called a *smart label* because it contains more information and has more capabilities than the barcode label. For example, with a smart label attached to it, an object can be tracked by an automated RFID system without manual intervention. By tracking the object globally using the EPCglobal network, you can automate its flow through the supply chain. This will result in supply chains that can tune themselves automatically to respond efficiently to the changing demands of the consumer: a big win. Supply chain is just one of a multitude of RFID applications with advantages of the same magnitude.

As an example, Figure 9.2 shows a roll of smart labels. Like a label, they can be conveniently printed on and attached to an object, and they have all the capabilities of an RFID tag—the best of both worlds. These labels were created on an RFID printer.

Figure 9.2 A Sample Roll of Smart Labels *(Image Courtesy of Weber Marking Systems Inc.)*

Working with RFID Printers

An RFID printer, also called a *smart label printer*, prints, well, smart labels. In this section, we will explore what a smart label printer is and how you install, configure, and troubleshoot it.

NOTE

The word *periphery* has its origin in a Greek word that means *circumference* or *outer surface.* So periphery, in general, means boundary or outer part of something: body or space. In the computer industry, a peripheral (also called a *peripheral device*) is a device that is not required for the computer to function, but it is used to expand the capabilities of the computer. For example, a printer attached to a computer is a peripheral device.

Understanding RFID Printers

An RFID printer is an RFID peripheral device that can print human-readable information on the surface of a smart label and can write data to the transponder (tag) inside the label. It's also called a *smart label printer* or *RFID printer/encoder*. In other words, the RFID printer combines the functionalities of a traditional barcode (or label) printer and an RFID encoder to create (print and encode) smart labels.

So, a smart label can contain UPC barcode information printed on its front surface as well as the information written (encoded) to an RFID tag sandwiched between its outer printable layers. Typical RFID printers encode data onto the HF (13.56 MHz RFID) and UHF (915 MHz RFID) tags. These printers can be connected to a PC via a port (parallel, serial, or USB) or through a network connection such as Ethernet and used as output devices. Typical RFID printers offer the following functionality:

- Print text on the surface of a smart label, e.g., destination address
- Print linear and 2-D barcodes on the surface of a label, which will be scannable
- Print graphic images on the surface of a label
- Write (encode) information to the tag inside a label
- Read the encoded information
- Verify the accuracy of the information
- Mark a faulty label and proceed to the next label

From now on in our discussion, by *label* we mean smart label, unless specified otherwise. Figure 9.3 presents Weber's R110XiIIIPlus RFID smart label printer from Zebra Technologies, an example of a smart label printer that produces smart labels compliant with Electronic Product Code (EPC) Gen 1 and Gen 2 protocols.

Figure 9.3 An Example of an RFID Printer *(Image Courtesy of Weber Marking Systems Inc.)*

To give you a real feel for typical characteristics of RFID printers, Table 9.1 presents values for some characteristics of the actual printer shown in Figure 9.3.

Table 9.1 Some Characteristics of the RFID Printer Shown in Figure 9.3

Printer Characteristic	Value
Maximum label length	39 inches
Maximum print width	4.0 inches
Label width	0.79–4.5 inches
Label depth	0.25–39.00 inches
Printhead density (print resolution)	300 dpi
Maximum print speed	10 inches per second
Tag frequency	UHF (915 MHz)
Memory	4 MB Flash, 16 MB SDRAM, Compact Flash (optional) up to 1 GB
Support for RFID protocols	EPC Gen1 and Gen 2
Weight	50 lbs.

NOTE

Label width and depth depend on the RFID inlay (tag) inside the label.

You will find a whole range of RFID printers in the market with a wide spectrum of features and capabilities. You will obviously select the one that is optimal for your application. Following are the typical requirements that you should consider in selecting a printer:

- **Accuracy verification** Seriously consider whether the printer has a feature to automatically void the faulty label, e.g., the label that has the transponder that does not respond properly to the read/write instructions from the printer.
- **Compliance** Ensure that the printer meets compliance requirements. For example, an RFID printer to be used for passive UHF EPC smart labels must have been tested to be EPC compliant.
- **Environmental condition** Consider the environmental condition such as temperature in which the printer will be used. This could help you determine the features, such as temperatures in which the printer can operate safely.
- **Flexibility** Consider the flexibility the printer provides in terms of properties such as the following:
  - Does it support various label sizes?
  - Does it support multiple tag protocols?
  - Does it let you configure the number of attempts it will make to write to the tag? This will let you have control over the time it will spend on faulty tags and the volume of error messages it will generate as a result.
- **Quantity of labels** Consider the volume of labels that you will be printing in the short and long term. This might help you determine features such as printing speed.
- **Software support** Ensure that the printer has the required software support.

EXERCISE 9.1

You have an RFID printer with the resolution of 300 dpi. How many dots can it print in one mm?

Solution:

1 inch = 2.5 cm = 25 mm
300 dpi = 300 dots/inch = 300 dots/25 mm = 12 dots/mm

Once you have selected and purchased an RFID printer suitable for your needs, it's time to install it.

Installing the RFID Printer

Although details about installing a printer can vary from printer to printer, the overall process is the same for all RFID printers. The main tasks in installing the printer are unpacking the printer, placing it firmly at a selected site, making all the connections, loading the printer with media and ribbon, and executing any installation software.

CAUTION

Install the label stock between the printhead and the platen before closing the pivoting deck; otherwise, the debris on the platen may damage the printhead. Also, do not touch the printhead or the electronic components under the printhead assembly.

You can install an RFID printer by following the instructions in the manual that will come with the printer. However, the main steps are listed in the following as a high-level view of the installation procedure:

1. **Unpack and check the content** Make sure all the listed items are present in the package.
2. **Select a printer site** The printer must rest safely on a solid, leveled surface, and there must be enough access space for activities, such as proper ventilation and cooling, opening it to change the media, and so on. In selecting the printer space, you should also consider the interference from other RF devices, such as readers and cell phones.
3. **Connect the power cables**
    a. Make sure the printer's power switch is in the Off position.
    b. Attach the AC power cord to the AC power receptacle in the back of the printer.
    c. Attach the AC power cord to a grounded (three-prong) electrical outlet of the proper voltage.

4. **Load the printer** Load the printer with labels and ribbon. The instructions might be shown right on the printer panel.
5. **Power on the printer**
6. **Run the installation program** The installation software might come with the printer, or you might need to download it from a Web site. However, if you have thought through the various options, this part of the installation should not be difficult. Just follow the instructions from the software program and make your choices. You might need to choose port type and connection type between the computer and the printer. If your RFID system uses the network and you want to make the printer part of the network, connect it to the print server.

While handling the printer during or after installation, take the following precautions:

- During unpacking or handling, do not place the printer on its backside, because it may damage the printer interface connector.
- While setting up a printer, avoid touching the electrical connectors to prevent damage due to ESD. The ESD can damage or destroy the printhead or electronic components in the printer.
- Make sure the printer is properly grounded. Failure to do so could result in electric shock to the operator. To comply with international safety standards, printers are usually equipped with a three-pronged power cord; one of the three prongs is the ground prong. In this case, do not use adapter plugs, and do not remove the ground prong from the cable plug. If you do need an extension cord, make sure it uses a three-wire cable with a properly grounded plug.
- Verify the required voltage on the printer's model number label on the back of the printer.
- Never operate the printer on its side or upside down.
- If you are using direct thermal mode, clean the platen roller, printhead, and lower and upper media sensors every time you change the media.

Smart label printers usually use the following thermal printing techniques to print:

- **Thermal transfer printing** A printer using this technique prints on a medium such as paper by melting a coating of ribbon that will stay glued to the medium. This technique, by definition, requires a ribbon.
- **Direct thermal printing** A printer using this technique prints on a medium such as paper by selectively heating the coated thermochromic medium when the medium passes over the printer's thermal printhead. This technique does

not require a ribbon. *Thermochromism* is the property of a material to change color due to a change in temperature.

During or after installation, you will need to configure the printer to customize it according to your needs.

Configuring the RFID Printer

RFID printers typically offer a degree of flexibility by allowing you to configure the values for a set of properties. The exact set of configuration properties and the values of their range may change from printer to printer. But to give you a practical feel, we use a real printer (the SL5000e Smart Label RFID Thermal Printer from Printronix) as an example, to discuss the configuration properties that are typically available in most RFID printers. These properties include the following:

- **Label length** This is the length of the smart label. In most applications, the label length that you choose will match the physical label length—that is, the actual label length of the media installed.
- **Label width** This property specifies the width of the label.
- **Media-handling method** This property specifies how the printer will handle the media (label stock):
  - **Peel-off** In this method, the optional rewinder is installed, and it prints and peels die-cut labels from the liner without assistance. The printer waits for you to remove the label before it can move to print the next label; this is also called *on-demand printing.* The printer displays the "Remove Label" message to remind you to remove the label before it can print the next one.
  - **Tear-off** With this method, the printer positions the label over the tear bar after printing and waits for you to tear off the label before printing the next one; this is also called *on-demand printing.* The printer displays the "Remove Label" message to remind you to remove the label before the next one can be printed.
  - **Tear-off strip** With this method, the printer will print on the media and send it out the front until the print buffer is empty; then it positions the last label over the tear bar for removal.
  - **Cut** With this method, given that the optional media cutter is installed, it automatically cuts media after each label is printed, or it can cut after a specified number of labels have been printed using a software cut command.

- **Orientation** This property specifies the image orientation that will be used when printing the label.
- **Paper feed shift** This property represents the distance to advance a label or pull back when the tear-off strip, tear-off, peel-off, or cut media-handling option is enabled. In other words, it helps position the label.
- **Print intensity** This property specifies the level of thermal energy from the printhead to be used for the given type of media and ribbon installed. A large value for this property means more heat (thermal energy) will be applied for each dot, which in turn affects the print quality.
- **Print mode** This property specifies the type of printing to be performed:
  - **Transfer** Indicates thermal transfer printing mode that uses the heat-sensitive ribbon to perform heat-based printing.
  - **Direct** Indicates direct thermal printing mode (no ribbon) that uses special heat-sensitive media to perform heat-based printing.
- **Printing speed** This property specifies the speed in units of inches per second (ips) at which the media (label) passes through the printer when printing.
- **Save-config property** This property allows you to save up to eight unique configurations to meet different print job requirements.
- **Save-up config** This property allows you to specify that the factory configuration or any one of the eight possible saved configurations be used as the power-up configuration.

The range of values and the default value for each of these properties is shown in Table 9.2 for a specific printer as an example.

Table 9.2 Examples of Configuration Properties for an RFID Printer and Their Values

Configuration Property	Range of Values	Default Value
Label length	–15 to 15 inches	–3 inches
Label width	0.1–8.5 inches	—
Media-handling method	Tear-off strip, tear-off, peel-off, cut, and continuous	Tear-off strip
Orientation	Portrait, landscape, inv. portrait, and inv landscape	Portrait
Paper feed shift	–0.50–12.8 inches	0.00 inches

Continued

Table 9.2 continued Examples of Configuration Properties for an RFID Printer and Their Values

Configuration Property	Range of Values	Default Value
Print intensity	–15 to 15 inches	Transfer mode: –3 Direct thermal mode: 0
Print mode	Transfer, direct	Transfer
Printing speed	2–10 ips	6 ips
Save config	1–8	1
Save-up config	1–8	Factory

NOTE

The print intensity and the print speed must be compatible with the media and ribbon type to achieve the optimal print quality and barcode grades. Also, the maximum print speed depends on the maximum printer width and the print resolution (dots per inch).

So, you have installed and configured your printer and are ready to roll. However, keep in mind that errors and problems are part and parcel of life, even in the technology world. This is where troubleshooting skills come handy.

Troubleshooting the RFID Printer

The RFID encoder in an RFID printer can detect a number of errors. When one of these errors occurs, the RFID encoder instructs the printer to perform the currently selected error action and displays the appropriate error message on the control panel's LCD, as shown in Table 9.3. The error action is selected according to the value of the Error Handling parameter on the RFID Control menu, which you can set to one of the following values:

- **None** No specific action is taken. The printer will discard the failing label data and continue to the next smart label.
- **Overstrike (the default)** The unacceptable smart label will be printed with a grid or error message over the label. If the Label Retry Count is greater than zero, the same smart label will be tried over and over again until the label retry count is exhausted. In case of a failure, after the last try the error message will be printed if the Overstrike Style is set to Error Type Msg. These error messages are shown in Table 9.4. If the Overstrike Style is set to Grid, a grid pat-

tern will be printed instead of an error message. The failed label will not be reprinted.

- **Stop** The printer discards the failing label data, displays the error message "RFID Error: Check Media," and halts. The failed label is discarded, and reprinting of the labels must be initiated by the host. Once the error is cleared, the label with the failed tag moves forward until the next label is in the position to be printed.

So, if an RFID tag within a smart label is found unacceptable after performing a defined number of retries, one of these actions is performed.

Table 9.3 RFID Control Panel Error Messages

Error Message	Description
RFID Comm Err Check Cable	RFID error: communication cannot be established with the RFID encoder. Reader will be set to Disable in the RFID Control menu and the previous port settings will be restored.
RFID MAX RETRY Check System	Error Handling = Overstrike in the RFID Control menu, and the Label Retry count has been exhausted.
RFID TAG FAILED Check Media	Error Handling = Stop in the RFID Control menu, and the RFID encoder could not read the RFID tag.

The Overstrike is the default option for Error Handling mode. Overstrike printing error messages from the printing software are shown in Table 9.4. The *n* in these error messages represents a number code that identifies the area in the printer software where the failure occurred.

Table 9.4 Printed Overstrike Error Messages

Error Message	Description
Tag R/W Err *n* Check media	The printer software has attempted to write to or read from the RFID tag, but the RFID encoder has indicated that the tag could not be written to or read from.
Tag Comm Err *n* Check cable	The printer software has temporarily lost communication with the RFID encoder, or communication between the printer software and the RFID encoder was not synchronized and had to be forced.
Precheck Fail *n* Check media	The RFID tag was automatically failed since it did not contain the correct preprogrammed quality code. This failure occurs only when the Precheck Tags menu item is set to Enable.

You can press **PAUSE** to clear an RFID control panel error message and consult the troubleshooting section in the printer documentation. Some tips for troubleshooting an RFID encoder part of the printer are shown in Table 9.5 as an example.

Table 9.5 Troubleshooting the RFID Encoder

Problem	Solution
Inconsistent results	Ensure that the media is loaded correctly and passes smoothly over the antenna. Refer to Printer Setup documentation.
No communication between the printer part and the reader part of the RFID printer	1. Ensure that the serial interface adapter and the serial cable are plugged properly into the printer. Consult the printer installation documentation. 2. Ensure that the reader is set to Enable in the RFID Control menu. 3. Use the RFID Test option available in the RFID Control menu (Admin User enabled) to read and display the current RFID tag content. Consult the vendor documentation if necessary.
The RFID encoder works, but it does not meet expectations	Ensure that both Error Handling and Label Retry are set to the desired values in the RFID Control menu.
Tag failed	1. The label is possibly misaligned. Perform the Auto Calibrate procedure to ensure that the label is at top of form. Refer to the vendor documentation. 2. Ensure that you are using the correct media: smart labels with RFID tags located in the correct position. 3. The RFID tag is possibly defective. Try another tag. 4. Ensure that the application does not send too few or too many digits to the RFID tag.

CAUTION

Before performing any maintenance procedure, always disconnect the AC power cord from the printer or from the power outlet. Failure to remove power could result in injury to you, damage to equipment, or both. If applying the power is necessary at some step of the procedure, according to the documentation, always consider the fact that the power is on and proceed carefully.

A standard technique in troubleshooting, which also equally applies to printer problems, is the fault isolation technique. You improve your chances of identifying the cause of the problem by following this technique:

1. Collect information. Ask the user (reporting the problem) to describe the problem. Collect as much relevant information as you can.
2. Reproduce the problem. Verify the problem by running a diagnostic printer test in which you replicate the conditions reported by the user and reproduce the problem.
3. Look for the standard error messages to start troubleshooting.
4. If you cannot get an error message described in the troubleshooting table of the documentation, use the so-called half-split method to narrow down the problem area:
    a. Start at a general level and work your way down to details.
    b. Isolate the faults by narrowing down to half the remaining system at a time, until the final half is a field-replaceable unit or assembly.
5. Make one change at a time or one corrective action at a time. Test the printer operation after every corrective action.
6. Replace the defective unit or assembly. Do not attempt field repairs of electronic components or assemblies. Most electronic problems are corrected by replacing the printed circuit board assembly, sensor, or cable that causes the fault indication.
7. Install any part you replaced earlier to diagnose the problem and that was not found defective.
8. Test the printer again. Return the printer to normal operation when the reported symptoms disappear.

After you have created the smart labels with the RFID printer, you will need to apply those labels to the products that need to be tracked. You might also need some device that will give you feedback on how your RFID system is doing. All these devices are called *ancillary devices.*

Understanding Ancillary Devices and Concepts

Ancillary devices are the devices that might not be an essential part of a core system but are useful additions to it. RFID encoders and label applicators are examples of ancillary devices in an RFID system.

NOTE

The word *ancillary* means something of secondary importance; something subordinate to something else. For example, an instructor's manual is ancillary to the textbook.

Encoders and Label Applicators

RFID printer encoders are used to write information on a tag inside a smart label, and label applicators are used to apply smart labels to items that need to be tracked.

RFID Printer Encoders

An RFID printer encoder is a component of an RFID printer that writes data to and reads data from the tag inside a smart label being used as a media for the printer. We already covered this topic in the previous section. However, here are the important points to remember about a printer encoder:

- They typically operate in HF (13.56 MHz) and UHF (915 MHz).
- The behavior of the encoder in a printer can be controlled through the RFID control menu.
- An RFID printer encoder supports the following functionality:
  - Reading the data from the tag
  - Modifying, deleting, or writing the data to the tag
  - Verifying the accuracy of the data on the tag

The common faults detected and the errors displayed by an RFID printer encoder were presented in Table 9.3. Troubleshooting tips for an RFID printer encoder were presented in Table 9.4.

Once the smart labels have been printed by the RFID printer, they will to be attached to the items that need to be identified and tracked. This is where the automated label applicator enters the RFID story.

Automated Label Applicators

An automated label applicator is a labeling machine that is designed to automate the process of applying labels to products. These machines, widely used in manufacturing and food processing and distribution departments, can also be used for applying smart labels.

NOTE

Automated label applicators are capable of placing labels consistently in the same place on each item to the accuracy of 0.5 mm, and at high speed, up to 80m/minute or 20 packages per minute.

Automatic label applicators can be grouped broadly into two categories: pneumatic piston applicators and wipe-on applicators, as described in the following sections.

Pneumatic Piston Label Applicators

The pneumatic piston label applicator is a machine that has a piston to stop a product on a line for labeling and then applies the label to it using the tamp-down or blow-on technique. For that reason, this category is also called *tamp-blow applicators*. This type of label applicator is suitable under following conditions:

- You want to apply labels to the front, sides, or top of the product (or package), typically on an automated production line.
- The product does not have to be moving at a constant speed.

This is how it works:

1. A sensor detects the product.
2. The labeling machine activates a pneumatic piston.
3. The piston stops the product for as long as it takes to apply the label.
4. The label is served on a vacuum plate, which is moved by the pneumatic piston to the package.
5. The label is then applied by using one of the following methods:
   - The label is simply pressed against the product, called a *tamp-down action*.
   - The label is blown onto the product: the vacuum in the vacuum plate is replaced with air that provides the pressure to blow the label onto the package. In other words, the label is blown onto the package by a blast of air. The advantage of this method is that the vacuum plate does not actually touch the package: action at a distance.

Wipe-On Label Applicators

A wipe-on label applicator is a machine that performs pressure-sensitive labeling using a roller or a brush that wipes down the label on a package. This type of label applicator is suitable under the following conditions:

- You want to apply labels to the bottom, sides, or top of a product (or package), typically on an automated production line or by using a short length of conveyor.
- The product must be reliably moving at a constant speed.

This is how it works:

1. A sensor detects the product.
2. At a fixed time later, the labeling machine starts to issue the label.
3. The label is applied by tamping it down with the help of a foam roller.

To apply the labels consistently at the same place on each item, the side of the item that is to be labeled needs to be in the same position each time and must be reasonably flat. An example of a wipe-on label applicator, the Geset Alpha-V40, is shown in Figure 9.4.

Figure 9.4 An Example of a Wipe-On Label Applicator *(Image Courtesy of Weber Marking Systems Inc.)*

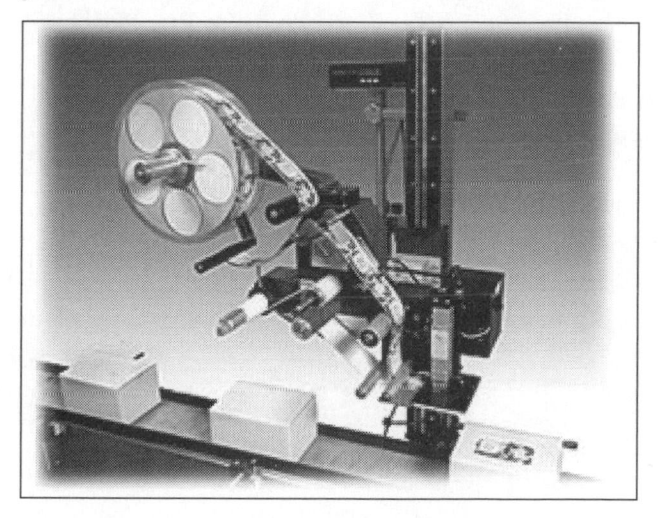

NOTE

There are many variants on the two types of label applicators that we have discussed. For example, in the case of pneumatic piston label applicators, a so-called 90-degree labeling machine can apply labels to the front face of a package (the leading face of a package in motion), and a labeling machine with twin vacuum plates can apply a label to two sides of a package, called round the corner. This kind of label applicator can also handle packages of variable sizes. A variant of a wipe-on label applicator can also apply labels to two sides of a product (round the corner).

In a nutshell, these two categories of label applicator use the following three label-placing techniques:

- The blow-on technique is used when the product surface must not be touched and when the accuracy of label placement is important.
- The tamp-down technique is suitable for pressure-sensitive labeling and uses a positive tamp action to ensure the complete label adhesion on the package.
- The wipe-on method is suitable for pressure-sensitive labeling when the product is moving at a reliably constant speed.

In fact, most automatic label applicators fall into these two categories and use one or more of these three techniques. However, the applicators come in different forms and shapes supporting different features, depending on the application requirements. Not all of them are fully automatic. So, there are other ways of categorizing the label applicators. Some of these are:

- **Automatic label applicators** These are the machines that are typically used on production lines. We have already discussed these under the wipe-on and pneumatic piston applicator categories. Automatic bottle labeling is another example.
- **Semi-automatic label applicators** In these applicators, part of the process is not automated. For example, the product might have to be manually staged for labeling. A variant of bottle labeling and handheld label applicators fall in this category. A handheld applicator is used to wipe the labels onto the products in a way similar to a pricing gun used in retail stores.
- **Print and apply label applicators** This is a system that offers an integrated solution by combining smart label printing and label applying functionalities. The label applicator part of the system can be either wipe-on or pneumatic piston, as described earlier.

An effective RFID system or network consists of diverse hardware, software, and logic to automate the identification and tracking process. An important component of automation is event-based action. For example, tell someone when a problem occurs, ask the reader to read when a tagged item shows up, and so on. This is where feedback systems come onto the scene.

Feedback Systems

Feedback systems in an RFID network, usually based on sensors, help the system run in an effective and automatic fashion. The input and output ports in an RFID system are used to integrate feedback with the system. For example, turning on or off lights or sound alerts could have a predetermined meaning as a feedback on the status of a process. For instance, warning noise and a red light might indicate missing items on a shipping pallet. Feedback systems, in general, are used for two purposes:

- Report problems with the system, e.g., to trigger a manual intervention.
- Perform an action within the RFID network based on an event—for example, to trigger the reader only when some tagged item shows up. This helps automate the RFID system.

Commonly used feedback systems or devices used in RFID systems or networks include photo eyes, light trees, horns, and motion sensors.

Photo Eyes

A photo eye, in general, is a sensor that detects the presence of something coming and reports the presence to another device. A photo eye works as an input device to an RFID system. It can be used as a proximity sensor that can provide presence and direction information of a load. For example, a photo eye sees an item coming on a conveyor and triggers the reader to begin reading. It can also be used to determine a good or bad read—that is, a good or bad tag. A bad read can prompt actions such as stopping the conveyor and triggering a light tree.

So, the following are some of the advantages of photo eyes in an RFID network:

- A photo eye can detect and tell the RFID portal when to start scanning a load passing along the conveyor. This minimizes the scan time and helps avoid interference between adjacent RFID readers.
- In conjunction with some business logic, a photo eye can also be used for verification of tags: they are properly encoded and placed.

Light Trees

A light tree refers to a stack of lights, connected to an output port of an RFID system, used as a feedback system. The light tree works as an output device because it reports some output of an RFID operation to the operator. It might be used to report a success, a failure, a warning, or information. A light tree is typically controlled by a software program called an *agent*. The agent forwards commands to the hardware to turn the tree lights on and off. As a practical example, in Websphere, IBM's software platform integrated with RFID, the following are the parameters of a light tree agent to dictate the behavior of the light tree:

- **duration.ms.beep** Represents the amount of time to signal when a beep request is received. The default value is 500 ms.
- **ignore.green.while.red** Dictates whether any green light indicators should be ignored if the light tree is currently red. The default value is false.

- **duration.ms.green** Represents the amount of time to signal when a green light request is received. The default value is 2000 ms.
- **duration.ms.red** Represents the amount of time to signal when a red light request is received. The default value is 2000 ms.

Table 9.6 presents some standard light signals, with examples, commonly used in RFID systems. A combined scenario using all three lights (green, yellow, and red) is presented in Exercise 9.2.

Table 9.6 Some Standard Light Signals Used in RFID Systems

Light	General Indicator	Example Scenario
Green or red	Success or failure	1. On a platform such as a pallet, when the containers are scanned, the system attempts to match the container tag with a purchase order in the database. If the scanned tag matches the purchase order, a green light displays on the light tree and the shipment can proceed; otherwise, a red light displays. 2. On a portal such as a dock door, a green light indicates that an expected item is received, whereas a red light indicates that an unexpected item is received.
Yellow	Warning	1. A yellow light can indicate that a worker is in an area where his safety will be compromised. 2. A yellow light can indicate improper temperature in the area where a reader is installed.
Flashing (blinking)	Attention	1. On a door at the back of a retail store, RFID readers and antennas are installed for receiving a shipment. As a box passes an RFID antenna, a flashing yellow light tells an operator that the tag on the box has been read. 2. An RFID reader reading tags on a pallet encounters an anomaly such as a faulty tag and triggers a flashing red light to tell the operator that a manual intervention is necessary.

EXERCISE 9.2

Consider a receiving station of a supply point where all the pallets are offloaded from a container and are put into the inventory of the supply point. An RFID-enabled forklift unloads the container and reads the tags on the pallets, and the application adds these pallets to the bulk storage inventory of this supply point. After offloading the container, the forklift takes each pallet to the distribution area. A door portal (connected to a host computer running a software application) has been set in the distribution area to ensure that no case has been removed from any pallet. The RFID system on the portal also has a light tree that consists of green, red, and yellow lights. How can this light tree be used as a feedback system for this application?

Solution:

- A yellow light can indicate that a case tag has been read.
- A green light can indicate that a pallet has been found intact; no case from this pallet is missing.
- A red light can indicate that something is missing from the pallet. In this case, a manual count could be performed.

The light tree uses light as a medium for output signals. There are output devices called *horns* that use sound as a medium for output (feedback) signals.

Horns

A horn is a feedback device that gets the attention of the operator by making a sound. It works as an output device in an RFID system. For example, the RFID system on a forklift may be equipped with an impact sensor. When the impact exceeds a specified value, the sensor reports it to the RFID system, which triggers the forklift's horn. Horns are generally used under extreme situations that require immediate attention, such as theft, intrusion, emergency, and so on.

Motion Sensors

A motion sensor is a device that detects the movement of an object in its surroundings by using some kind of waves such as ultrasonic waves, which are sound waves beyond the human ear's hearing capability. Such a sensor works as an input device to an RFID system. The wave emitted by the sensor is reflected by the object to the sensor, which then calculates the distance of the object from it. The changing distance (or position) of the object means the object is moving. Motion sensors can be used in an RFID system to indicate the arrival of tagged items at a portal such as a dock door. In this case, the RFID system, after getting the alert from the sensor, will trigger the reader to read the arrived items. The motion sensors can also be used against theft by combining them with the tags attached to the item that must not be moved. A motion of that tagged item will trigger the motion sensor, which in turn will trigger some alarm system.

So, feedback systems and devices report problems with the RFID network and help further automate the RFID system by facilitating event-based action. *Event-based action* means an action performed when a certain event occurs. For example, start a read cycle if an item is detected in a certain area. Such automation gives rise to powerful RFID systems such as real-time location systems.

Real-Time Location Systems

A real-time location system (RTLS) is an RFID system that automatically and continuously tracks and reports in real time the location of assets and personnel that are tagged to be tracked. An RTLS designed to track moving objects typically works as follows:

- **Moving tags** An item that needs to be tracked is tagged typically with an active tag that regularly transmits signals at some time intervals. The tag, of course, contains the unique identification code for the item.
- **Stationary readers** At certain locations, stationary readers are waiting for the tagged items to pass. When a tagged item passes through this location, the readers receive the signal from the tag because the tag is continuously transmitting. Upon receiving a signal from the tag, they collect the information from the tag about the item, and they determine the location of the item. The readers send the location and the item data to a monitoring system.
- **Monitoring system** The monitoring system receives the location information about an item being tracked from the readers.

CAUTION

A wide variety of RTLS can be designed. Depending on the application requirements, an RTLS can use both active and passive tags and can be used to track both moving and stationary objects.

Note the difference between a usual RFID portal and an RTLS. In a typical RFID portal, the tags are read when they pass the portal as part of a structured process, whereas RTLS tags are read automatically and regularly, independent of the process that moves the tags. With RTLS, no intervention or controlled process is needed to determine asset location; it is automated.

RTLS can continuously determine and track the real-time location of moving assets and personnel using active RFID tags attached to moving objects. Another type of RTLS, also called *local locating system (LLS),* can be designed to track objects in a constrained location (indoor or outdoor) for tracking assets within a corporate facility. In this case, readers are installed at key locations within a facility. Active tags attached to objects typically broadcast their identity at regular intervals.

NOTE

In an RTLS, a tag is generally read by multiple readers installed at different points in a location. Each reader can determine its distance from the object by the time it took to get the response for its signal sent to the tag on the object. By putting together the distances of the object from all the readers and the locations of all the readers, the system can accurately calculate the exact location of the object. If the number of readers involved is three, this technique is called the *triangulation technique*.

RTLS are getting popular in many fields, including healthcare, manufacturing, and logistics, where they help locate and manage high-value assets in daily operations. Following are some examples of the use and advantages of RLTS:

- Efficiently and automatically identify and track valuable assets to ensure that these assets remain in the facility. This helps reduce theft.
- Track patients and doctors within a hospital facility. Without an RTLS, finding doctors, nurses, or patients at critical times without delay could be a challenge.
- RTLS can also help address patient safety and security issues. For example, an RTLS can be used to trigger an alarm if the wearer of a tag leaves a certain perimeter, the authorized or safe area.
- RTLS can help you save time that you would otherwise waste in looking for things. RTLS will tell you exactly where the things are.

The three most important takeaways from this chapter are the following:

- RFID printers help integrate the barcode technology with RFID technology as they are used to print smart labels like barcode labels and write data to and read data from the tags that exist inside the smart labels.
- An automatic label applicator applies labels to products, typically on production lines, in an automatic fashion using one of three labeling techniques: blow-on, temp-down, or wipe-on.
- Photo eyes and motion sensors are input feedback devices, whereas light trees and horns are output feedback devices.

Summary

Smart labels represent the junction between two technologies: barcode and RFID. Smart labels are created using an RFID peripheral device called an *RFID printer*, which can print human-readable information on the surface of a label and can write data to the transponder (tag) inside the label. Furthermore, an RFID printer can verify the correctness of the data on the tag and can void the smart labels with faulty tags inside them. Smart labels, created by an RFID printer, can be applied by automatic label applicators to the items that need to be tracked. An automatic label applicator applies labels to products, typically on production lines, in an automatic fashion using one of three labeling techniques: blow-on, temp-down, or wipe-on. In the tamp-down technique, the label is simply pressed against the item; in a wipe-on technique, the label is tamped down with the help of a foam roller; and in the blow-on technique, the label is blown onto the item by a blast of air.

Once you have deployed an RFID system, including peripheral devices, you always need to monitor the system in operation. There will be operational and performance issues and problems with the system and its components. How can you troubleshoot problems with an RFID system? The next chapter explores this topic.

Exam's Eye View

Comprehend

- ☑ An RFID printer is an output device that is connected to a computer. It can print human-readable information on the surface of a smart label and can write data to the transponder (tag) inside the label.
- ☑ An RFID printer encoder is a component of an RFID printer that writes data to and reads data from the tag inside a smart label being used as media for the printer.
- ☑ An automated label applicator is a machine that is used to apply smart labels to items in an automatic fashion. It can be integrated with a smart label printer.
- ☑ A photo eye and motion sensors are input devices, whereas a printer, light tree, and horn are output devices.

Look Out

- ☑ Smart labels offer backward compatibility with barcode technology; that is, they can work with the applications that use barcode technology in an old-fashioned way.

☑ A ribbon is required if the printer is configured to print in transfer thermal mode, whereas no ribbon is required for direct thermal mode.

Memorize

☑ RFID printers can print information on the surface of a label and can read information from the tag inside the label as well as write information to it.

☑ RFID printers typically support smart labels with HF (13.56 MHz) and UHF (915 MHz) tags.

☑ The width and depth of a smart label depend on the size of the tag inside.

☑ An RFID printer, on average, takes about 2 seconds to print one label.

☑ Automated label applicators are capable of placing labels consistently in the same place on each item to the accuracy of 0.5 mm and at high speed, up to 80 m/minute or 20 packages per minute.

☑ There are three main labeling techniques:

☑ The blow-on technique is used when the product surface must not be touched and when the accuracy of label placement is important.

☑ The tamp-down technique is suitable for pressure-sensitive labeling; it uses a positive tamp action to ensure the complete label adhesion on the package.

☑ The wipe-on method is suitable for pressure-sensitive labeling when the product is moving at a reliably constant speed.

Key Terms

Automated label applicator A labeling machine that is designed to automate the process of applying labels to products.

Barcode A specific precise arrangement of parallel lines (bars) and spaces that vary in width to represent a specific piece of data. It's a machine-readable representation of information printed on a surface in a visual format.

Direct thermal printer A printer that produces a printed image using heat-sensitive paper when the paper passes over a thermal printhead.

Horn A feedback device that gets the attention of the operator by making a sound. It works as an output device in an RFID system.

Light tree A stack of lights, connected to an output port of an RFID system, used as a feedback system. The light tree works as an output device because it reports some output of an RFID operation to the operator.

Logistics A technique for obtaining resources such as products, services, and people when and where they are needed. It's the science of the process to get the right quantity at the right time for the right price.

Motion sensor A device that detects the movement of an object in its surroundings using some kind of waves such as ultrasonic waves, which are sound waves beyond the human ear's hearing capability. This works as an input device to an RFID system.

Photo eye A sensor that detects the presence of something coming and reports this presence to another device. It works as an input device to an RFID system.

RFID printer An RFID peripheral device that can print human-readable information on the surface of a label and can write data to the transponder (tag) inside the label. It's also called a *smart label printer* or an *RFID printer/encoder.*

RFID printer encoder A component of an RFID printer that writes data to and reads data from the tag inside a smart label being used as media for the printer.

Smart label A barcode label that has an embedded RFID tag inside it. You can print human-readable, useful information on the label face, such as sender's address, destination address, and product information.

Thermal head A component of a thermal printer that generates heat to print on paper.

Platen A rubber roller that feeds paper into a printer.

Thermal transfer printer A nonimpact printer that uses heat to register an impression on paper via a heat-sensitive ribbon.

Thermochromism The property of a material to change color due to a change in temperature.

Universal product code (UPC) A data structure (an encoding scheme) originally used to write barcodes, widely used in retail stores in the United States and Canada.

Real-time location system (RTLS) An RFID system that automatically tracks and reports in real time the location of assets and personnel that are tagged to be tracked.

Self Test

A Quick Answer Key follows the Self Test questions. For complete answers and explanations to the Self Test questions in this chapter as well as the other chapters in this book, see **Appendix A**.

1. Which two of the following are not ancillary input devices for an RFID system?

   A. A motion sensor

   B. Photo eyes

   C. Light trees

   D. Horns

2. You need to track a cow when it passes through certain locations of a pasture. You have installed readers at those locations but no motion sensors. The readers are always powered on, but they are not continuously scanning. What kind of tag will you use to tag the cow?

   A. Active tag

   B. Passive tag

   C. Semipassive tag

   D. No need for a tag

3. The feedback input devices such as a photo eye in an RFID network typically are:

   A. Readers

   B. Tags

   C. Sensors

   D. Host computers

4. To create smart labels, an RFID printer is usually connected to a:

   A. RFID reader

   B. Computer

   C. Tag

   D. EPCglobal network

5. All of the following are the label-placing techniques used by automatic label applicators except:

   A. Tamp-down

   B. Wipe-on

   C. Blow-on

   D. Direct thermal

6. Match the items in the second column of the table to the items in the first column.

Device	Function
A. Light tree	1. An input device that can sense something coming to a reader portal such as a dock door
B. Motion sensor	2. In conjunction with software, it can help determine bad tags or badly placed tags
C. Photo eye	3. An output device that can be used to signal information or a warning to the operator
D. Printer encoder	4. Helps writes data to a tag inside a smart label.
E. RFID printer	5. Prints smart labels

Self Test Quick Answer Key

For complete answers and explanations to the Self Test questions in this chapter as well as the other chapters in this book, see **Appendix A**.

1. C and D
2. A
3. C
4. B
5. D
6. A-3
   B-1
   C-2
   D-4
   E-5

Chapter 10

RFID+

Monitoring and Troubleshooting RFID Systems

Exam Objective	What It Really Means
2.1 Given a scenario, troubleshoot RF interrogation zones (e.g., root-cause analysis).	Understand root-cause analysis and how to monitor and troubleshoot an interrogation zone. You should understand various kinds of monitoring: intrusive, nonintrusive, status, and performance. You must understand that variation in the values of some performance metrics could indicate instability of or problems with the system. You must also understand the common reasons for hardware failures and the standard troubleshooting procedure.
2.2 Identify reasons for tag failure. 2.3 Given a scenario, contrast actual tag data to expected tag data.	Understand how to monitor andtroubleshoot tag failures. You must know how to identify improperly tagged items, and you must understand the reasons for tag failures. You should also know how to manage tag failures.

Introduction

After you have designed and installed your RFID system, what will inevitably come next? The answer to this question is problems, all kinds of problems, including reader failure, tag failure, network connectivity problems, and so on. This is where monitoring and troubleshooting enter the scene.

An RFID system is made of three components: hardware, software, and data. The hardware consists of tags, readers, and antennas. The software largely consists of middleware that manages to take the data from the reader to the enterprise applications; it comprises the nervous system of the EPCglobal network. Monitoring your RFID system involves status monitoring and performance monitoring. Performance monitoring involves measuring some performance metrics. To find the causes for the problems identified during monitoring, you will need to troubleshoot the system.

So, the core question addressed in this chapter is how to monitor and troubleshoot an RFID system. In search of an answer, we will explore three avenues: monitoring, measuring performance metrics, and troubleshooting.

Monitoring an RFID System

Monitoring any system is essential to foreseeing and detecting problems and keeping the system running effectively. Because RFID is an evolving technology, you might find a lack of literature available to guide you in monitoring and troubleshooting an RFID system. However, most of the important research and development in any discipline occurs by people applying analogies to the way similar problems were solved in related disciplines and building on experience acquired in the same discipline. RFID is a closely related field to IT, if not a part of it.

The first lesson we can apply from the IT world is that RFID systems should support remote monitoring and troubleshooting of equipment, to improve efficiency and effectiveness. System manageability requires that good RFID solutions incorporate remote management, which will include remotely monitoring equipment, remotely identifying problems and their causes, providing alerts for problems such as failures, and supporting both problem troubleshooting and problem resolution.

So, the monitoring techniques used in IT and other fields can be applied in some way to RFID. In this section we will discuss some of those techniques in the context of RFID. A very important problem-solving methodology used in many fields, including IT, is *root-cause analysis.*

Understanding Root-Cause Analysis

Organizations often have problems that afflict their operations and result in reduced profits and increased customer dissatisfaction. Quite often, organizations try to fix a problem quickly by addressing the symptoms on the surface and without ever finding what really caused the problem in the first place. This approach causes the problems to reoccur over and over again. The goal of *root-cause analysis (RCA)* is to prevent the problem from reoccurring by finding and eliminating the root causes. In other words, RCA is the application of a set of problem-solving methods based on identifying and eliminating (or correcting) the root causes of a problem. The fundamental assumption in the RCA approach is that the problems are best solved by correcting or eliminating the root causes, as opposed to merely addressing the immediate, obvious symptoms that appear on the surface.

TIP

By applying corrective measures to root causes once, you must not expect that the problems will disappear forever, even though that is the goal. You can only hope that the likelihood of problem recurrence will be minimized, because completely preventing a problem from reoccuring through a one-time intervention is not always possible. Look at RCA as an iterative process and a tool for continuous improvement.

RCA involves different approaches from a variety of fields for getting at the root causes of a problem. The main approaches are:

- **Barrier analysis** According to this approach, the root causes of a problem exist in failed or missing barriers such as safety measures, unrecognized risks, or inadequate safety engineering. The process of finding the root causes includes assessing the adequacy of deployed barriers or recognizing the lack of barriers. This approach, which has its origin in the fields of accident investigation and occupational safety and health, is also called *safety-based RCA*.
- **Change analysis** According to this approach, the root cause of a problem often exists at the level of strategic management and organizational culture. To find the root cause, compare the process with the unsuccessful outcome to a similar process with a successful outcome. This approach has its origin in the fields of change management, risk management, and system management and is also called *system-based RCA*.
- **Events and causal factors analysis** According to this approach, the root causes of a problem lie in not conforming to one or more of the steps in a process. This is similar to the malfunctioning of one step out of a set of sequential steps in a production line. Finding the root causes here includes examining the events, related conditions, and causal factors in chronological order. This approach has its origin in the fields of quality control for industrial manufacturing and is also called *production-based RCA*.

- **Tree-diagram analysis** This approach, an extension of the events and causal factor analysis, assumes that the root causes of a problem lie in the process failure in general. The underlying philosophy of this approach is that you can eliminate the recurrence of problems and increase performance by improving processes. Finding the root causes using this approach includes creating and using tree diagrams describing the factors contributing to the event that caused the problem. This approach is also called *process-based RCA*.

Depending on your organization, your system, and the problem, you can determine which of these RCA approaches is suitable for you. However, the general high-level process for performing root-cause analysis is the same:

1. Identify and clearly define the problem.
2. Gather the relevant data, including evidence.
3. Identify the causal factors: all the causes or problems that contributed to the occurrence of this problem.
4. Find root causes for each causal factor, and iteratively arrive at the ultimate root cause of the problem.
5. Develop recommendations for solutions.
6. Implement the approved solutions.

So, keep the following principles of RCA in mind:

- Applying corrective measures to root causes of a problem is more effective than just addressing the symptoms of the problem.
- To get optimal results, RCA must be performed in a systematic fashion, and its conclusions must be backed up by evidence.
- There can be (and usually is) more than one root cause for a given problem.

So far, we have discussed RCA in general. But how does it apply to an RFID system? RFID is an evolving part of the IT industry. In the IT industry, it is a common practice to look at various data inputs from related incidents and identify the root causes by recognizing the correlations between those incidents. The same practice is being applied in its basic form to RFID systems.

The important point here is that in system problem solving, you need to go beyond just core monitoring of individual devices and collect some data that will help troubleshoot the problem. Is this data available from normal operation, or do you collect it just for troubleshooting the problem? This brings us to a discussion of the various types of monitoring.

Understanding Monitoring

To monitor a system, you need information (data) about the system. For example, to monitor an RFID system, you will need information from readers. Depending on how you receive this information, there are two kinds of monitoring:

- **Nonintrusive monitoring** In this type of monitoring, you use the information that is available from normal operation. Therefore, you place no extra requirement for information on the RFID system. For example, all the information that a reader gathers from normal interrogation of a tag falls into this category.
- **Intrusive monitoring** This is a type of monitoring that requires collecting the information that is not available from normal operation of the system. For example, your reader might offer commands that you can issue to gather information about its internal status or condition. This kind of intrusive information can help you detect (or predict) problems that might not be evident from nonintrusive information.

Usually nonintrusive information is enough to find the status of a device, such as a reader—for example, whether it's working or not. However, nonintrusive information might be needed to look deeply into a problem and find the root cause.

NOTE

Nonintrusive monitoring might be good enough to check the status of a reader, whereas intrusive monitoring might be needed to get a deeper insight into the system, to find the root cause of a problem or to predict a future problem.

From the perspective of the system details that you want to monitor, there are two kinds of monitoring: status monitoring and performance monitoring.

Status Monitoring

Status monitoring consists of monitoring the basic status of the system and the devices in the system, such as the following:

- Is the device connected to the network?
- Are the devices powered?
- Are the antennas operating?
- Is a reader reading the tags successfully?

Some solutions include a status indicator panel on your desktop that will allow you to monitor the status of all readers from one location. Feedback systems such as light tree and horns, discussed in Chapter 9, can also play an important role in this type of monitoring.

The next level of detail in monitoring is performance monitoring.

Performance Monitoring

Performance monitoring consists of monitoring the performance of a system and the devices in the system, such as the following:

- Read rates of readers
- Reading accuracy
- Error frequency: how frequently an error occurs

The performance data on predetermined metrics such as read rates helps determine the normal behavior of the system and identify the variance in the normal behavior and hence a problem. For this reason, performance monitoring is also called *behavior monitoring*.

Equipped with this understanding of monitoring your RFID system, you are ready to explore ways to troubleshoot an interrogation zone.

Monitoring and Troubleshooting Interrogation Zones

You often monitor a system by monitoring a set of metrics. A *metric* is an observable property that can be measured. It's also called a *quantity*. A metric is composed of some parameters that define the system and its performance. Following are the parameters that characterize an interrogation zone:

- Reader failure
- Number of tags passing through an interrogation zone
- Number of tags being successfully read
- Number of tags that are not being successfully read (read errors)

Based on these parameters, you can define metrics that can be measured or calculated based on the measurements and that will indicate the status of an RFID system and its performance. Some examples of such metrics are discussed in the following sections.

Mean Time Between Failures (MTBF)

Mean time between failures (MTBF) is the average time between two consecutive failures of a device or a system. Usually there is an underlying assumption in calculating MTBF:

After each failure, the system is fixed and returned to service immediately. This is a measure of reliability, robustness, and stability of the system. It can be applied to system components or to the system itself. In our case, the system is the RFID system and the components are the antennas, readers, host computer, and other network elements.

What do we mean by a failure? We need to define that. For example, on one extreme, you can consider that a reader has failed if it's not operational at all; on the other end of the spectrum, you could consider it failed when it misses a tag read or creates read error. In general, you can determine a threshold of read errors, and when the number of read errors exceeds the threshold, the reader may be considered failed.

MTBF indicates the robustness of the system measured in the past and, based on that measurement, predicts the rate of failure in the near future. MTBF can be calculated using the following simple equation:

$$MTBF = T_L/N_F$$

where:

- *TL* is the total lifetime (or operation time) of the device or the system over which the MTBF is being measured.
- *NF* is the total number of failures recorded.

For example, if a reader fails twice during 200 hours of operation, the MTBF can be calculated as in the following:

$$MTBF = T_L/N_F = 200/2 = 100 \text{ hours}$$

A more involved example is presented in Exercise 10.1.

Exercise 10.1

You have an RFID system with 100 readers. You have collected reader failure data for 200 hours. During this time, 10 readers failed once and another set of five readers failed twice (each of the five readers failed twice). What is the MTBF of this system?

Solution:

T_L = 100 readers x 200 hours = 20,000 reader hours

N_F = 10 + 5 x 2 = 20 reader failures

MTBF = T_L/N_F = 20,000 reader hours / 20 reader failures = 1000 hours/failure

A reader reads tags. So, tag traffic rate is obviously another metric of interest.

Average Tag Traffic Volume

Average tag traffic volume (ATTV) is the average number of tags passing through an interrogation zone during an interval of time. The interval can be a minute, 10 minutes, an hour, or whatever you determine it to be. This metric is important for the following two reasons:

- It indicates how much load the reader has to deal with on a portal.
- When a reader reads a tag, it typically sends the collected data to a host computer. So the tag traffic increases the data traffic in the network.

To measure ATTV, the monitoring system collects the following data from the readers:

- Tag counts
- The time at which the tag was counted

With this data, the ATTV can be calculated using the following equation:

$$ATTV = 1/N\ S^{i=N}{}_{i=1}\ t_i$$

where:

- *N* is the number of intervals for which the measurement is being taken.
- *ti* is the number of tags counted during an interval denoted by *i*.

Let's work through a simple example. Assume that you determine the interval to be 15 minutes, and you collect the data over an hour. In the four quarters of the hour, 50, 35, 30, and 45 tags are detected, respectively. So, you can calculate the ATTV as shown in the following:

$$ATTV = 1/N \sum^{i=N}{}_{i=1} t_i = 1/4\ (50 + 35 + 30 + 45) = 160/4 = 40 \text{ tags per 15 minutes}$$

How do you determine the time interval? That depends on your application and the system requirements. But keep in mind that if your interval is too long, say, a few hours, you cannot see the pattern of traffic (how the traffic changes during this interval), and if your pattern is too short, you'll have too many data points to deal with unnecessarily.

So, the ATTV indicates the following:

- How much tag traffic is flowing through the interrogation zone
- The pattern of traffic, that is, how the traffic changes with time; for example, you can see the pattern by taking 24 measurements of ATTV each day

From the ATTV measurements, you can predict how much traffic is expected to pass through an interrogation zone during a certain period. The accuracy of this prediction partly depends on the amount of data that was collected to make this prediction, that is, the statistical uncertainty.

Actual Versus Predicted Traffic Rate

It's always of interest to measure the variance between the predicted values of a metric with its actual value. Actual versus predicted traffic rate (APTR) is the variance of the actual tag traffic from the predicted tag traffic through an interrogation zone over a time period. The predicted tag traffic rate can come from the ATTV measurements in the past. A significant variance of actual traffic rate from the predicted traffic rate could indicate a problem with the system.

The APTR can be calculated using the following equation:

$$APTR = 1/N \sum_{i=1}^{i=N} (t_a - t_p)$$

where:

- N is the number of intervals for which the actual measurement is taken.
- *ta* is the actual current ATTV.
- *tp* is the predicted ATTV from the past measurements.

The larger the magnitude of APTR, the larger the variance of the actual value from the predicted value, and therefore the louder is the alarm that there could be something wrong with the system. Make sure you are comparing the actual value to the predicted value for the same time interval, because the predicted (and also the actual) values for different time intervals could be different. For example, in a given day, there might be more tag traffic from 2:00 to 3:00 P.M. than from 7:00 to 8:00 A.M.

Read Errors to Total Reads Rate

Read errors to total reads rate (RETR) is the total number of read errors divided by the total number of read attempts. A *read error* is reader's a failed attempt to read a tag. The measure of RETR can indicate the problems that caused the read errors, including:

- A faulty antenna
- Faulty tags or improperly tagged items
- Improper placement of antennas
- Low signal strength
- Signal interference, signal absorption, or any other adverse environmental effect

To measure RETR, the monitoring system collects the data about number of read errors, number of successful reads, and the time interval during which these read attempts were made. With this data collected for a few intervals, you can calculate RETR using the following equation:

$$RETR = (\sum_{i=1}^{i=N} E_i) / (\sum_{i=1}^{i=N} E_i + \sum_{i=1}^{i=N} S_i) = 1 / (1 + \sum_{i=1}^{i=N} S_i / \sum_{i=1}^{i=N} E_i)$$

where:

- *N* is the number of intervals for which the measurements are taken.
- *Ei* is the number of read errors during the interval *i*.
- *Si* is the number of successful reads during the interval *i*.

A high value for RETR should be taken as an alarm for a problem with the RFID system (or portal): either an internal problem or due to adverse environmental effects such as absorption and interference.

The value of RETR can change over time.

Read Error Change Rate

Read error change rate (RECR) is the variance of RETR over time. It indicates the instability or unreliability of the RFID system. For example, a continuous increase or fluctuation in the value of RETR indicates an underlying problem with the system. Upon troubleshooting, you might find a fault with the design of the system or with the hardware components.

EXERCISE 10.2

Table 10.1 shows the number of tags counted by a reader in its interrogation zone.

Table 10.1 Reader Tag Count for Exercise 10.2

Time Interval	Tag Counts in the Interrogation Zone
9:00–9:10	45
9:10–9:20	50
9:20–9:30	35
9:30–9:40	60
9:40–9:50	55
9:50–10:00	31

During this time:
Total number of read errors = 50
Total number of successful reads = 600
Calculate ATTV and RETR.

Solution:

$ATTV = 1/N \sum^{i=N}_{i=1} t_i = (1/6)(45 + 50 + 35 + 60 + 55 + 31)$
$= 276/6 = 46$ tags per 10 minutes

$\sum^{i=N}_{i=1} S_i / \sum^{i=N}_{i=1} E_i = 600/50 = 12$

$RETR = 1 / (1 + 12) = 1/13 = 1/13$

Or simply:

$RETR = 50 / (50 + 600) = 50/650 = 1/13$

A reader reads tags in its interrogation zone. Readability problems can also occur due to faulty tags.

Monitoring and Troubleshooting Tags

Even though the process of manufacturing of tags and their application to items has significantly matured, tag failures do still occur due to various reasons. Therefore, monitoring and troubleshooting tags are important tasks for an RFID professional. First of all, you need to ensure that the tags are properly placed on items. Furthermore, you need to know the reasons the tags can fail and how to manage tag failures. The tags are placed on items to be tracked before the items get out into the world. So, the first step in monitoring tags is to identify improperly tagged items.

Identifying Improperly Tagged Items

Proper tag placement on items that need to be tagged is crucial to the success of an RFID application. The challenge is to choose the right spot on an item where the tag will be placed. The right spot really depends on the kind of item; it will vary from one type of item to another. Therefore, testing to choose the right spot is necessary before you start using label applicators to place labels on a mass scale.

Even after you have chosen the right spot to place tags on items, there will be misplaced or improperly placed tags, which will cause problems. So, it's important to identify those tags. There are four kinds of improperly tagged items:

- Items that are tagged with faulty tags.
- Items on which the tags are placed incorrectly, where they cannot be read properly by the reader.
- Items on which the tags are placed at the correct spots but on which the tags are applied incorrectly, perhaps bent or folded.
- Items on which the tags are not properly oriented when correct orientation is required for efficient reading.

Tags can fail due to the defects introduced during manufacturing, and manufacturing defect rates are relatively high in smart labels, discussed in Chapter 9. If you are using smart label printers to print these labels, it will identify the smart label with a faulty tag and mark it void. For pre-encoded smart labels, you will lose the identification numbers (serial numbers) along with the faulty label.

Following are some of the methods for identifying improperly tagged items:

- Inventory discrepancies can indicate improperly tagged items because the tags on the items were not properly read; hence the items were not counted.
- Some feedback devices, such as photo eyes and motion sensors, provide the presence information of an item and instruct the reader to read it. If the reader cannot see it, it could be an improperly tagged item.
- Improperly tagged items can also be identified by combining the information from the sensors with the logic in the software.
- Improperly tagged items can be identified during automatic application of labels by placing an interrogation portal right after the application point—for example, on a read portal. If the tag cannot be read, it might be defective or improperly placed.
- You can also identify improperly tagged items via manual inspection, perhaps before shipping.

The tags on improperly tagged items are prime for failure. There are, however, a multitude of reasons that a tag would fail.

Identifying Reasons for Tag Failures

A tag failure can be defined as the inability of a properly functioning reader to detect a tag when it's scanning its interrogation zone. In other words, a tag is considered failed when a reader cannot detect it when the tag is within the reader's read range. By identifying and understanding the reasons for tag failures, you can take steps to avoid failures, identify failures, and fix problems, thereby optimizing the performance of your RFID system. Following are the main reasons for tag failures:

- **Manufacturing defects** These are defects that are introduced during manufacturing. They can be detected in the process of testing or applying the tag. For example, if you are using a smart label printer to print these labels, it will identify the smart label that has a faulty tag and mark it void. The printer can also introduce the error because it can write to the tag inside the smart label.
- **Wrong tag type** A tag failure can happen because a wrong tag type was applied for a given application.

- **ESD** The ESD, discussed in Chapter 8, can damage the transistors in a tag's IC and thereby cause the tag to malfunction.
- **Harsh environmental conditions** Harsh environmental conditions can cause adverse effects that can affect tag detection in the following ways:
  - Some conditions such as extreme temperature can damage tags, causing them to fail.
  - Some materials in the environment can cause effects such as interference, reflection, diffraction, and scattering that can prevent the reader from properly communicating with the tag, making it look as though the tag does not exist.
  - Tags can also be damaged by mishandling of the tagged items.
  - The material of the package surface or the content can affect the tag read. For example, the read rates could be very low for tags placed on metallic containers due to reflection and interference problems.
- **Improper placement** An improperly placed tag on an item can be damaged or simply might not be detected by the reader. Improper placement includes the wrong spot on the item for the tag; the wrong way to place the tag, such as folding it; or the wrong orientation of the tag.
- **Dense tag environment** The dense tag environment, discussed in Chapter 4, also prevents a tag from being detected by a reader. For example, the shadowing effect created by the dense tag environment prevents a tag from being read by the reader.

To fix the problems caused by tag failures and to control the damage, you need to manage tag failures.

Managing Tag Failures

Tag failures can have a significant impact on the effectiveness and efficiency of an RFID system. Therefore, it is very important to manage tag failures. You can improve system performance by identifying and possibly fixing tag failures before the tagged items get into the tracking network. In general, you need to manage tag failure before applying the tags, during the tag application process, and after the tag application, when the item is being tracked.

Management Prior to Applying Tags

Some integrated automatic tag application solutions offer the following failed tag management features:

- Identify and eliminate the faulty tags prior to applying them to the items.

- Track the statistics on the failed tags and communicate them to the RFID validation stations.

Management during Application

You can set up a test interrogation portal (zone) immediately after the point where the tags are being applied by an automatic applicator. The idea behind managing the tag failure prior to and during tag application is to prevent these tags entering the tracking system. You will certainly improve system performance by identifying and fixing tag failure problems before they enter the tracking system. However, tag failures do occur when the items are being tracked, and you need to manage those failures as well.

Management after Applying the Tags/During Tracking

The items with failed tags that are being tracked will generate data inconsistencies such as inventory discrepancies. To deal with such problems, it is important that tag failures are identified and the information reported to the application program or the database management system automatically. The EPCglobal network helps in management of failed tag data in the tracking system by making the information about the globally tracked items visible throughout the tracking system, such as supply chain. This network consists of five elements:

- **EPC** The identification number on the tag that uniquely identifies the tagged item.
- **EPC tags and readers** The reader reads the tag of an item being tracked and sends its EPC to a host computer or application system running object naming service (ONS).
- **ONS** A mechanism that uses domain name system (DNS) on the Internet to discover the information about an object that has been tagged with an EPC number (unique ID). ONS itself does not contain the information about the object, but it knows where the information is; for example, it knows the IP addresses of the servers that have the information.
- **Physical markup language (PML)** A language that is used to write the data (information) about the object in the format that is convenient for communication. Once the information about an object is found, that needs to be communicated. PML is used to store and communicate that information. Because this information should be available throughout the EPCglobal network, we need a distributed data management system. That's where Savant enters the picture.
- **Savant** A specification developed by the Auto-ID center at MIT for a software system in the middle of data sources such as readers and enterprise appli-

cations. It provides a distributed data management system that manages the data for the objects being tracked in the EPCglobal network. Application-level events (ALE) is the new name of the game in this field.

NOTE

Details about Savant and ALE are outside the scope of this work. However, you should know why ALE is known as the specification for application event management systems. An *event* is defined as some occurrence of interest in a device or a system. For example, a successful read of a tag by a reader is an event, and so is a read error. Managing the data from these events is called *event management*.

In the context of managing the elements of an RFID system (e.g., tags and readers), you should also be aware of Simple Network Management Protocol (SNMP), which is used to remotely manage devices connected to a TCP/IP network and is based on two components, agent and manager. An *agent*, which is housed in a managed object, gathers information about the object (device). A *manager* sends requests to the agent to get information about the object and to execute commands on the object.

Once you find an item with a failed tag, what would you do? Depending on the situation, here are some of the solutions:

- Use a backup such as a barcode when you print the smart labels. This will come in handy if the tag in the smart label fails.
- Fix the tag, if it's fixable.
- Replace the tag with another tag.

An RFID system is composed of three elements: hardware, software, and data. Readers and tags can be considered hardware components that have data. There are some general techniques to monitor and troubleshoot hardware failures, which we discuss next.

Monitoring and Troubleshooting Hardware

A *hardware failure* is the failure of a hardware component to function to its specifications. As an RFID professional, you should understand the causes of hardware failures and know the tips and techniques for diagnosing and troubleshooting hardware problems.

Understanding the Causes of Hardware Failures

An RFID system can experience hardware failures for a number of reasons, including the following:

- **Damaged hardware components** Following are some of the reasons for this damage:
  - Harsh environmental conditions such as extreme temperatures can damage hardware components.
  - Unregulated power supply of any sort can damage hardware components. This includes power surges from the regular power line or from lightning.
  - ESD and improper grounding of equipment can also contribute to the failure of hardware components.
- **Incompatible hardware components** This could include readers and tags using different communication protocols or methods.
- **Unperformed reader firmware upgrades** Firmware upgrades were discussed in Chapter 3. The basic idea is that the readers' firmware must be compatible with the protocols being used by the tags that the reader is trying to read. The protocols are still evolving, so it's very important that you buy readers whose firmware is upgradeable.

TIP

Before starting to troubleshoot it, restart a misbehaving hardware component such as a reader to see if that simple action solves the problem. However, consider the consequences and impact of restarting a component of a running system. Thinking through a proposed solution is always a good idea.

Diagnosing RFID Hardware Failures

Following are some tips to diagnose and troubleshoot hardware failures:

- **Change** Find out what has changed since the last time a failing component was working properly. The change could include the installation of new software or a new hardware component in the system. This step is always a good start in troubleshooting.
- **Verification** Sometimes within a system, it's not obvious that it is a given hardware component that is causing the problem. In that case you need to

verify the failure of that hardware component. One way of doing that is to replace the suspect component with a component that has just been tested and found working. If after replacement the problem goes away, you can conclude that the replaced component is failing.

- **Start simple** Try the simplest thing first. For example, make sure the component is powered on if it is supposed to be, and inspect the cable connections—for example, between the interrogator and the antenna and between the interrogator and the network, and so on.
- **Network reachability** To check the network connection and reachability, you can use the *ping* command, which is part of TCP/IP.

CAUTION

When you are replacing or repairing a hardware component, take safety precautions. For example, make sure the power to the component is turned off.

Sometimes it's obvious that a hardware component has failed. Other times it's not, and you simply conclude from a problem that the system is experiencing a hardware component failure. In this case, you need to troubleshoot the problem.

Standard Troubleshooting Procedure

The following standard procedure works well for both hardware and software troubleshooting:

1. **Identify and define the problem** You identify and define the problem clearly by performing the following steps:
   a. **Establish symptoms** You do this by observing the problem and collecting information about the problem.
   b. **Identify the affected area** You need to identify which area of the system is affected by the problem.

   If you have identified and defined the problem, in most cases you should be able to reproduce it.
2. **Identify causes of the problem** Once you have identified and defined the problem, the next step is to find what is causing the problem. Remember, the cause of a problem could be another problem. The first step toward finding the cause is to ask what has changed since the last time the system was working fine. Use the method of elimination to narrow down and isolate the real cause.

This involves eliminating relatively obvious causes or problems and making your way to more complex causes or problems.

3. **Select the most probable cause** Select the most probable cause and work on it. It's important to work on one cause at a time, and make one change at a time, else you will not be able to map the effects with the cause.
4. **Plan and implement a solution** This can involve repair or replacement of a component.
5. **Test the implementation** After you have implemented a solution, you should test it and see if it solves the problem. Note and recognize the effects of the solution. Sometimes a solution can cause other problems.
6. **Document the solution** You must document your solution. This step is very important but is often ignored. Having good documentation will save you a lot of time and effort if the same problem appears again in the future. It will also be useful for the next RFID professional who will take your place, in case you move on.

It's usually a good practice to replace a *field replaceable unit (FRU)* instead of trying to repair it. An FRU is a component of a system that a user or technician can quickly and easily remove from the system and replace without having to send the entire system to a repair facility.

The three most important takeaways from this chapter are the following:

- Status monitoring consists of monitoring the status of a system and its devices (e.g., whether a device is powered on); performance monitoring consists of measuring the performance metrics of a system or a device.
- By measuring the system performance metrics, you determine the normal values of these metrics. Significant variation in the values of some of these metrics can indicate instability of or problems with the system.
- The standard troubleshooting procedure is identify the problem, identify the cause, implement the solution, test the solution, and document the solution.

Summary

Monitoring and troubleshooting are essential parts of running an effective RFID system. Usually nonintrusive information (collected during the normal operation of a system) is enough to find the status of a device, whereas intrusive information (collected as part of monitoring but not available as part of normal operation) might be needed to help you look deeply into the problem and find its root cause. You can find the root cause of a problem using root-cause analysis (RCA), which is the application of a set of problem-solving methods based on identifying and eliminating (or correcting) the root causes of a problem. Problems are identified during monitoring via either status monitoring or performance monitoring. Status monitoring consists of monitoring the basic status of the system and its devices, such as whether the reader is powered on; performance monitoring consists of monitoring the performance of a system and the devices in the system via measuring performance metrics such as such as readers' read rates. By measuring the system performance metrics, you determine the normal values of these metrics; significant variation in the values of some of these metrics could indicate instability of or problems with the system. The causes of a problem can be found during troubleshooting, which has the following standard steps: identify the problem, identify the cause, implement the solution, test the solution, and document the solution.

Exam's Eye View

Comprehend

- Nonintrusive monitoring might be good enough to check the status of a reader, whereas intrusive monitoring might be needed to get a deeper insight into the system, to find the root cause of a problem or to predict a future problem.
- Mean time between failures (MTBF) is a measure of reliability, robustness, and stability of a system and can be applied to a system or its components.
- Average tag traffic volume (ATTV) shows the pattern (changing volume) of tag traffic through an interrogation zone.
- A tag failure is defined as the inability of a properly functioning reader to detect a tag when it's scanning its interrogation zone. So, a tag is considered failed when a reader cannot detect it when it is within read range.

Look Out

- Root-cause analysis is an iterative process; applying corrective measures to the root causes of a problem once does not guarantee that the problem will not reoccur.
- MTBF indicates the robusteness of the system measured in the past and, based on that measurement, predicts the rate of failure in the near future.
- A significant variance of actual value of a metric such as tag traffic rate from the predicted value based on the past measurements could indicate a problem with the system.
- A high value for read errors to total read rate (RETR) should be taken as an alarm for a problem with the RFID system (or portal)—either an internal problem or one due to adverse environmental effects such as absorption and interference.
- Read error change rate (RECR) is the variance of RETR over time and indicates the instability or unreliability of the RFID system.
- Items with failed tags that are being tracked will generate data inconsistencies such as inventory discrepancies.

Memorize

- MTBF = TL/NF
- ATTV = $1/N \sum_{i=1}^{i=N} t_i$
- APTR = $1/N \sum_{i=1}^{i=N} (t_a - t_p)$
- RETR = $1/(1 + \sum_{i=1}^{i=N} S_i / \sum_{i=1}^{i=N} E_i)$

Key Terms

Actual versus predicted traffic rate (APTR) The variance of actual tag traffic from predicted tag traffic through an interrogation zone over a time period. The predicted tag traffic rate can come from the ATTV measurements in the past.

Application-level event (ALE) A specification, developed by EPCglobal, for RFID event management. It's a successor of Savant.

Average tag traffic volume (ATTV) The average number of tags passing through an interrogation zone during an interval of time.

Field replaceable unit (FRU) A component of a system that can be quickly and easily removed from the system and replaced by the user or a technician without having to send the entire system to a repair facility.

Mean time between failures (MTBF) The average time between two consecutive failures of a device or a system.

Metric An observable property that can be measured. It's also called a *quantity*.

Object naming service (ONS) A mechanism that uses domain name system (DNS) on the Internet to discover information about an object that has been tagged with an EPC number (unique ID).

Physical markup language (PML) A language that is used to write data (information) about an object in a format that is convenient for communication.

Root-cause analysis (RCA) The application of a set of problem-solving methods based on identifying and eliminating or correcting the root causes of problems, with the goal of preventing the problem from reoccurring.

Read error A reader's failed attempt to read a tag.

Read error change rate (RECR) This is the variance of RETR over time. It indicates the instability or unreliability of an RFID system.

Read error to total reads rate (RETR) The total number of read errors divided by the total number of read attempts.

Savant A specifciation, developed by the Auto-ID center at MIT, for a distributed middle software system between data sources in the EPCgloabl network, such as readers and enterprise applications, with a goal of filtering the data, such as eliminating duplication. Savant was designed to work as a distributed data management system for the EPCglobal network. It was a predecessor of ALE.

Simple Network Management Protocol (SNMP) A TCP/IP protocol used to remotely manage devices connected to a TCP/IP network or the Internet.

Tag failure The inability of a properly functioning reader to detect a tag when it's scanning its interrogation zone.

Self Test

A Quick Answer Key follows the Self Test questions. For complete answers and explanations to the Self Test questions in this chapter as well as the other chapters in this book, see Appendix A.

1. Which two of the following is not true about root cause analysis?

   A. Applying corrective measures to root causes of a problem is more effective than just addressing the symptoms of the problem.

   B. To get optimal results, RCA must be performed in a systematic fashion, and its conclusions must be backed up by evidences.

   C. There can be (and usually is) more than one root cause for a given problem.

   D. You only need to perform the rot cause analysis once to prevent a problem from reoccurring.

2. All of the following are the reasons for the RFID tag failures except:

   A. Wrong tag type

   B. Manufacturing defect

   C. A tag was read by more than one in the same interrogation zone

   D. ESD

3. Which of the following is not a method to identify improperly placed tags on items?

   A. Use a feedback device such as a photo eye

   B. Vary the power emission by the reader

   C. Set up an interrogation zone right after the point of applying the tags on a production line

   D. Manual inspection

4. A host computer and a reader are connected to the same network. You want to test the network reachability of the reader. Which of the following command will you issue on the host computer?

   A. ipconfig

   B. reboot

   C. testreach

   D. ping

5. To find the root cause of an RFID system failure:
   A. Nonintrusive monitoring is enough
   B. You may need intrusive monitoring
   C. Status monitoring should be enough
   D. Performance monitoring is not needed
6. The ping command is a part of:
   A. MTBF
   B. EPCglobal standards
   C. TCP/IP
   D. Reader firmware
7. The ping command is a part of:
   A. MTBF
   B. EPCglobal standards
   C. TCP/IP
   D. Reader firmware
8. The read error rate may indicate any of the following except:
   A. Faulty antenna
   B. High power emission by antenna
   C. Improper placement of antennas
   D. Low signal strength
9. Which of the following two measures the stability of an RFID system? (Choose two.)
   A. Changing read error rate
   B. Average tag traffic volume through an interrogation zone
   C. Mean time between failures
   D. Read error rate
   E. Operating frequency

10. An RFID system can experience hardware failure due to any of the following except:

    A. Damaged reader

    B. Intrusive monitoring

    C. Incompatible reader and tags

    D. Reader's firmware not supporting protocol used by tags

Self Test Quick Answer Key

For complete answers and explanations to the Self Test questions in this chapter as well as the other chapters in this book, see the Self Test Appendix.

1. D
2. C
3. B
4. D
5. B
6. C
7. C
8. B
9. A and C
10. B

Glossary

Active tag. A tag that has its own power source such as a battery, and can initiate the communication by sending its own signal.

Actual versus predicted traffic rate (APTR). The variance of the actual tag traffic from the predicted tag traffic through an interrogation zone over a time period. The predicted tag traffic rate can come from the ATTV measurements in the past.

Adjacent channel interference. The interference from a signal with the frequency close to the operating frequency of the RFID system.

Air interface protocols. Set of protocols that define the rules for communication between tags and readers.

Ambient EM noise (AEN). The EM noise existing at a given location such as a compartment, a room, or a particular outdoor location.

Antenna gain. Ratio of energy radiated at a point of maximum radiation from an antenna to the energy radiated at the same point by some reference antenna.

Antenna. The device used to transmit and receive signals such as radio waves. Both a reader and a tag have their own antennas through which they communicate with each other.

Application level event (ALE). A specification, developed by EPCgloabl, for the RFID event management. It's a successor of Savant.

Attenuation. Means decrease in the amount of something. In RF physics, it means the decrease in amplitude (strength) of the RF signal (wave).

Automated label applicator. A labeling machine that is designed to automate the process of applying labels to products.

Average tag traffic volume (ATTV). The average number of tags passing through the interrogation zone during an interval of time.

Backscattering. The process of collecting inbound signal (energy), changing the signal (the data it carries), and reflecting it back to where it came from.

Band congestion interference. The interference resulting from the overcrowding of frequency bands, that is, too many devices operating within a shared frequency band or closely spaced frequency bands.

Bandwidth. The width of a band of electromagnetic frequencies used for transferring data. It is a measure of how fast data can be transferred on a given transmission path and determines the total data transmission rate that the path can offer. The basic units of bandwidth are HZ (cycles/sec) in the analogous world, and bits/sec in the digital world.

Barcode. A specific precise arrangement of parallel lines (bars) and spaces that vary in width to represent a specific piece of data. It's a machine-readable representation of information printed on a surface in a visual format.

Beacon. A tag that emits a signal at predetermined intervals. Beacons are mostly used in real time locating systems (RTLS).

Beamwidth. The angle between the two half-power points around the point (the main lobe) that has the peak effective radiated power.

Bistatic antenna configuration. A configuration in which an interrogator will use one antenna for sending signals and another antenna for receiving signals. This configuration enables the full duplex communication mode.

Blueprint. A site diagram used to visualize the site infrastructure. A blueprint, in general, is any plan that documents an architecture or an engineering design.

Cable loss. The amount of signal power lost in the cable being used as a transmission line.

Carrier signal. The wave that carries the data signal.

Characteristic impedance. The impedance of the transmission line when it's assumed lossless and of infinite length.

Choke point. A specific location through which lots of items pass. The interrogation zones are usually set up at choke points.

Circularly polarized antenna. An antenna that radiates circularly polarized wave, which is a wave in which the electric field vector rotates in a circle as the wave travels.

Corrosion. The deterioration of essential properties of a material object due to reactions with its environment. A common example of corrosion is rust.

Data signal. The wave that actually contains the information that needs to go to the receiver.

Data Transmission Rate. Actual rate, in bits/sec, at which the data is being transmitted over a communication line from one device to another. The transmission can be wireless as well.

Dense interrogator environment. An area in which multiple interrogators are operating in close proximity to one another.

Dense tag environment. An area in which multiple tags are in the interrogation zone of an interrogator so that more than one tag can get the same signal from the interrogator.

DHCP. Dynamic host configuration protocol used to automatically (dynamically) provide IP addresses to the devices connected to the network.

Diffraction. Refers to the bending of EM wave when it strikes the sharp edges or when it passes through a narrow gap (slit).

Dipole antenna. An antenna that consists of a straight electric conductor, made of conducting metal such as copper, interrupted at the center, and therefore making two poles.

Direct thermal printer. A printer that produces a printed image by using heat sensitive paper when the paper passes over the thermal print head.

Directivity. The ability of an antenna to focus in a particular direction to transmit or receive energy. It is calculated as the ratio of the maximum value of power transmitted (or received) per unit solid angle to the average power transmitted (or received) per unit solid angle.

Effective radiated power. The power that will need to be supplied to a reference antenna in order to produce the same power as this antenna is radiating in a specific direction.

Electronic Product Code (EPC). A group of coding schemes for tags defined by the standard called Generation 2.

Electrostatic discharge (ESD). The instantaneous electric current created by the flow of electrons from high density (of electrons) surface to low density surface, for example, when the two surfaces rub against each other.

Electrostatic discharge (ESD). The instantaneous electric current created by the flow of electrons from high density (of electrons) surface to low density surface, for example, when the two surfaces rub against each other.

Environmental interference. The interference form natural sources of EM radiations such as lightening and solar radiation.

EPC number. A number assigned by the manufacturer to identify the specific object to which the tag is attached. This is also called object ID (OID).

EPC. The electronic product code, a family of coding schemes for Gen 2 tags.

EPCglobal Network. A set of RFID technologies that enables immediate automatic identification and sharing of information on items in the supply chain.

EPCglobal. A joint venture between GS1 (formerly known as EAN International) and GS1 US™ (formerly the Uniform Code Council, Inc.) created to commercialize the EPC technology which was originally developed at the Auto ID center at MIT.

Far-field. The EM radiations beyond the antenna's near-field. In the far-filed, the signal power decreases as square of the distance from the antenna.

Federal Communications Commission (FCC). An independent United States government agency established by the Communications Act of 1934 as the successor to the Federal Radio Commission and is charged with regulating all non-Federal Government use of the radio spectrum (including radio and television broadcasting), and all interstate telecommunications (wire, satellite and cable) as well as all international communications that originate or terminate in the United States.

Field Replaceable Unit (**FRU)** A component of a system that can be quickly and easily removed from the system and replaced by the user or the technician without having to send the entire system to a repair facility.

Firmware. A software program embedded in a device that configures its basic functionality when the device is powered up.

Full duplex. This is the mode in which data can be transferred between two devices in both directions simultaneously.

Full Faraday Cycle Analysis (FFCA). A process to collect data regarding the EM waves in a site environment over a full business cycle which is typically 24 to 48 hours. Business cycle in this case is the time that includes all the normal operations involving the frequency band in which you are collecting the data.

Ground loop. An unwanted current that flows in a conductor connecting two points that are supposed to be at the same potential, e.g. ground (or zero) potential, but are actually at different potentials.

Ground. Electrical connection to earth. The part directly in contact with the earth is called the earth electrode, and it can be as simple as a metal rod or a wire. For example, multiple ground connection paths between two components of an electrical system will create a ground loop.

Half duplex. This is the mode in which data transfer between two devices can only occur in one direction at a time.

High frequency (HF). The frequency band of 3-30 MHz. The RFID systems in this band operate at 13.56 MHz.

Horizontal polarization. Linear polarization in which the wave travels horizontal to the surface of the Earth.

Horn. A feedback device that gets the attention of the operator by making sound. It works as an output device in an RFID system.

HTTP. Hypertext transfer protocol, a protocol on which the World Wide Web is based. Web browsers and web servers use this protocol to interact with each other.

ICMP. Internet control message protocol, used by the routers on the Internet to report errors in communication.

IEC. The International Electrotechnical Commission (IEC), an international standards organization in the area of electrical, electronic and related technologies. Some of its standards are developed jointly with ISO.

Impedance. Resistance to the flow of current in a circuit element, and is measured as a ratio of voltage across the element and current through the element.

Inlay. The combination of antenna, chip, and substrate.

Insert. An inlay inserted between a label in the front and an adhesive layer in the back. The adhesive in the back can be permanently mounted, for example, on the inner wall of a tire.

Institute of Electrical and Electronics Engineers (**IEEE**). An international non-profit, professional organization for the advancement of technology related to electricity and electronics. There are about 900 active IEEE standards.

Interference. The interaction between two waves. The signal wave can interact with other waves that it meets on the way to its destination. A resultant wave is produced as a result of interference and the receiver receives the resultant wave.

Interference. The interaction between two waves. The signal wave can interact with other waves that it meets on the way to its destination. A resultant wave is produced as a result of interference and the receiver receives the resultant wave.

Interoperability. The ability of systems or components of a system to provide services to and accept services from other systems or components, and thereby operate together effectively to provide services to the user.

Interrogation zone. The area around an interrogator within which it can successfully communicate with a tag. In other words, if a tag enters the interrogation zone, it can be interrogated by the interrogator.

Interrogator. The RFID component that collects information from tags and sends it to a host system. The process of collecting the information from the tags is called reading the tags, and for this reason an interrogator is also called a reader.

IP. Internet protocol, used to define IP addresses for devices and to send data to communicate with a destination device.

ISM (industrial, scientific, and medical). A group of frequencies, originally reserved for non commercial uses in industrial, scientific, and medical fields, now generally used for the RFID systems.

ISO (International Organization for Standardization). An international standards body composed of representatives from national standards bodies. Founded on February 23, 1947, this organization sets world-wide industrial and commercial standards, which are popularly called ISO standards.

Isotropic antenna. A hypothetical non-directional antenna that radiates power uniformly in all directions. It is often used as a reference to calculate quantities such as effective radiated power (ERP).

ITU (International Telecommunication Union). An international organization established to standardize and regulate international radio and telecommunications. It was originally founded with the name *International Telegraph Union* in Paris on May 17, 1865.

Jamming. The interference or the noise caused by an intentional emission of radiation by another device or system.

Kill function. Used to disable the tag permanently.

Label. A tag attached to an item with an adhesive with the purpose of identifying the item.

Light tree. A stack of lights, connected to an output port of an RFID system, used as a feedback system. The light tree works as an output device because it reports some output of an RFID operation to the operator.

Linearly polarized antenna. An antenna that radiates linearly polarized wave, which is a wave in which the electric field vector stays parallel to a line in space as the wave travels.

Link margin. Refers to the ratio of maximum effective signal strength received to the minimum signal strength received. In RFID, it means the amount of power that a tag can extract from the RF signal before the communication between the tag and the reader weakens.

Logistics. A tool for obtaining resources such as products, services, and people, when and where they are needed. It's the science of the process to get the right quantity at the right time for the right price.

Low frequency (LF). The frequency band of 30-300 KHz. The RFID systems in this band typically operate at 125 KHz or 134 KHz.

Mean time between failures (MTBF). The average time between two consecutive failures of a device or a system.

Metric. A metric is an observable or property that can be measured. It's also called a quantity.

Microwave frequency. A frequency band of 1 GHz-300 GHz. The RFID systems in this band typically operate at 2.44 GHz or 5.80 GHz.

Modulation. The process that encodes the data signal into the carrier signal and creates the radio wave that is actually transmitted by the antenna to propagate.

Monostatic antenna configuration. The configuration in which an interrogator uses the same (one) antenna for sending and receiving signals. This configuration only enables half duplex mode of communication.

Motion sensor. A device that detects the movement of an object in its surroundings by using some kind of waves such as ultrasonic waves, which are sound waves beyond human ear's hearing capability. This works as an input device to an RFID system.

National Electrical Manufacturers Association (NEMA). An organization that came into existence in1926 from the merger of the Electric Power Club and the Associated Manufacturers of Electrical Supplies, and provides a forum for the standardization of electrical equipment. It also helps the electrical industry by functioning as a central confidential agency for gathering, compiling, and analyzing market statistics and economics data.

Near-field. The EM radiations within the distance of the order of one wavelength from the antenna. In the near-filed, the signal power decreases as cube of the distance from the antenna.

Network connection. A connection made between two devices by connecting them to the same network through their network interfaces (cards). For a network connection, a device must have an IP address.

Noise. An unwanted electrical wave (or energy) present in a circuit or in a signal.

Noise. An unwanted electrical wave (or energy) present in a circuit or in a signal.

Object naming service (ONS). A mechanism that uses domain name system (DNS) on the Internet to discover the information about an object that has been tagged with an EPC number (unique ID).

Omnidirectional antenna. An omnidirectional antenna is a non-directional antenna that radiates power uniformly in all directions. An ideally perfect omnidirectionl antenna is also called isotropic antenna.

Operating frequency. The frequency of the radio waves that the interrogator and the tag use to communicate with each other.

Passive tag. A tag that does not have its own power source such as a battery and therefore cannot initiate the communication.

Path loss contour analysis (PLCA). The process of determining how the field strength and shape of the RF coverage in an interrogation zone varies. The PLCA data has the information about how the RF signals (waves) are degraded and distorted and how the wavefront (the shape of coverage) changes throughout the interrogation zone.

Path loss contour mapping (PLCM). The process of preparing a blueprint that maps the PLCA data.

Photo eye. A sensor that detects the presence of something coming and reports this presence to another device. It works as an input device to an RFID system.

Physical markup language (PML). A language that is used to write the data (information) about the object in the format that is convenient for communication.

Platen. A rubber roller that feeds paper in a printer.

Polarization. Refers to the direction of oscillations in the EM waves transmitted by the antenna.

Polarization. The property of EM waves such as RF waves that determines the direction of the electric field in the plane perpendicular to the direction of wave propagation.

Power over Ethernet (POE). A system that transmits electrical power, along with the data, for example, to remote devices over the standard twisted-pair cable in an Ethernet network.

Power supply unit (PSU). A device that supplies electrical energy to an output load or group of loads.

Read cycle. A scan for RFID tags performed by a reader. The reader can run read cycles periodically or on demand.

Read error change rate (RECR). This is the variance of RETR over time. It indicates the instability or unreliability of the RFID system.

Read error. A failed attempt by a reader to read a tag.

Read errors to total reads rate (RETR). The total number of read errors divided by the total number of read attempts.

Read range. The maximum distance from which a tag can be read.

Read rate. The number of tags that a reader can read per unit time. Sometimes, read rate is also used for maximum data transfer rate, that is, the maximum rate at which data can be read from a tag, expressed in units of bits/sec.

Reader collision. An interference in communication because two or more interrogation zones are overlapping. This situation is a result of a dense interrogator environment.

Reader speed. The speed with which a reader collects information. It can vary from about 25 tags/sec to over 1000 tags/sec.

Reader. The RFID component that collects information from tags and sends it to a host system. The process of collecting the information from the tags is called reading the tags, and it is also called interrogating the tag. For this reason, a reader is also called an interrogator.

Reading efficiency. The ratio of number of successful reads to the total number of read attempts.

Reading speed. The number of tags a reader can read per unit time.

Real time location system (RTLS). An RFID system that automatically tracks and reports in real time the location of assets and personnel that are tagged to be tracked.

Reflection. The abrupt change in direction of a wave at an interface between two dissimilar media so that the wave returns into the medium from which it hit the interface.

Refraction. The change in direction of a wave at an interface between two dissimilar media, but the wave does not return back to the medium from which it hit the interface.

Regulation. A legal restriction promulgated by a government administrative agency through rulemaking and is typically supported by a threat of consequences such as fine for not following it.

Relative humidity. A quantity used to describe the amount of water vapor that exists in a gaseous mixture of air and water.

Resonance The characteristic of a system to absorb more energy when the frequency of its oscillations matches the system's natural frequency (resonant frequency) than it does at other frequencies.

RFID portal. The area where RFID tags can be read or written to.

RFID printer encoder. A component of an RFID printer that writes data to and reads data from the tag inside a smart label being used as a media for the printer.

RFID printer. An RFID peripheral device that can print human readable information on the surface of the label, and can also write data to the transponder (tag) inside the label. It's also called smart label printer or RFID printer/encoder

Root cause analysis (RCA). An application of a set of problem solving methods based on identifying and eliminating or correcting the root causes of the problems with the goal of preventing the problem from reoccurring.

SAR. Specific absorption rate, a measure of the rate of energy absorbed by (or dissipated in) an incremental mass contained in a volume element of dielectric materials such as biological tissues.

Savant. A specification, developed by Auto-ID center at MIT, for the distributed middle software system in between the data sources in the EPCglobal network such as readers and the enterprise applications with the goal of filtering the data such as eliminating duplication. It was designed to works a distributed data management system for the EPCglobal network. It was a predecessor of ALE.

Scattering. The phenomenon of absorbing a wave and re-radiating it, and thereby changing its direction.

Semi-passive tag. A tag that has its own power source such as a battery, but does not initiate the communication.

Serial communication. Process of transferring data from one device to another sequentially one bit at a time.

Serial connection. A connection set up between two devices by connecting their serial ports through a cable.

Simple network management protocol (SNMP). A TCP/IP protocol used to remotely manage devices connected to the TCP/IP network (or internet).

Smart label. A barcode label that has an embedded RFID tag inside it. You can print human readable useful information on the label face such as sender's address, destination address, and product information.

Spectrum analyzer. A device used to examine the spectral composition of an EM wave. You can sue it to measure the signal strength, interference, and AEN.

Spurious emissions. The interfering radiations transmitted outside the operating frequency band in the form of narrowband signals or wideband noise.

Standard. Guideline documentation (specifications) that reflects agreements on products, practices, or operations by nationally or internationally recognized industrial, professional, trade associations or governmental bodies. If all the vendors follow the same standard, the product from those vendors will be compatible with each other and will be interoperable.

Standing wave. A pattern of waves produced out of the interference of two waves of the same frequency traveling in opposite directions on the same transmission line.

Stray tag read. A read of a tag, which is not supposed to be read by a reader. For example, due high power, a reader can read tags outside its planned interrogation zone.

Substrate. A support structure (layer) that houses a tag's antenna and chip.

Tag collision. An interference in communication that occurs because two or more tags try to respond to an interrogator at the same time. This situation is a result of dense tag environment.

Tag failure. The inability of a properly functioning reader to detect a tag when it's scanning its interrogation zone.

Tag identifier (TID). A code assigned by the manufacturer to uniquely identify the product.

Tag starvation. A situation created by Aloha based anti-collision protocols in which a tag has to wait for long time before it can be identified by a reader.

Tag. An RFID component attached to an item that needs to be tracked. It contains the information about the item and provides that information on request.

TCP. Transmission control protocol, used for reliable communication with a specific application on a destination device.

Thermal head. A component of a thermal printer that generates heat to prints on paper.

Thermal transfer printer. A non-impact printer that uses heat to register an impression on paper by using a heat sensitive ribbon.

Thermochromism. The property of a material to change color due to a change in temperature.

UDP. User datagram protocol, used for simple but unreliable communication with applications on other devices.

Ultrahigh frequency (UHF). A frequency band of 300 MHz-3GHz. The RFID devices in this band operate at different frequencies in different regions of the world.

Uninterruptible power supply (UPS). A device that maintains a continuous supply of electric power to the connected equipment by supplying power from a battery when the utility power becomes unavailable.

Universal product code (UPC). A data structure (an encoding scheme) originally used to write barcodes widely used in U.S.A. and Canada in the retail stores.

Vertical polarization. Linear polarization in which the wave travels perpendicular to the surface of the Earth.

Voltage Standing Wave Ratio (VSWR). The ratio of maximum voltage to minimum voltage along the transmission line.

Wavefront. Refers to the geometrical shape of the space occupied by a traveling wave. For example an EM wave from an isotropic antenna travels in the free space in all directions making spherical wavefronts.

Appendix A

Answers to Chapter Self Tests

Learning Objectives

- **Understand why the correct answers are correct**
- **Understand why the incorrect answers are incorrect**

Chapter 1

Chapter 1 does not cover any exam objectives. Therefore, there are no review questions in this chapter.

Chapter 2

1. Answer: C

C is the correct answer because the wavelength depends on the velocity of the wave and its frequency, not on its amplitude.

A, B, and **D** are incorrect answers because they are the correct statements. Remember the formula ??= v/f, where *v* is the velocity of the wave in the transmission line and is less than the velocity of light *c*, which will be the velocity of the wave through the free space.

2. Answer: D

D is the correct answer because the signal-to-noise ratio is measured in decibels, denoted as dB.

A and **C** are incorrect because the signal-to-noise ratio is the ratio of the signal strength to the noise strength.

D is incorrect because ohm is not the unit used to measure resistance.

3. Answer: B

B is the correct answer because refraction causes the change in direction of a wave at an interface between two dissimilar media, but the wave does not return to the medium from which it hit the interface.

A is incorrect because reflection causes the abrupt change in direction of a wave at an interface between two dissimilar media so that the wave returns to the medium from which it hit the interfce.

C is incorrect because diffraction refers to the bending of an EM wave when it strikes sharp edges or when it passes through a narrow gap (slit).

D is incorrect because absorption refers to absorbing the energy of the wave by a medium and not bending.

4. Answer: D

D is the correct answer because amplification is not a modulation technique.

A, B, and **C** are incorrect answers because amplitude modulation, frequency modulation, and amplitude shift keying are all valid modulation techniques.

5. Answer: C

C is the correct answer because the VSWR relates to the phenomenon that occurs in the transmission line.

A, B, and **D** are incorrect answers because all these characteristics affect the propagation of RF waves through the free space.

6. Answer: A and B

A and **B** are the correct answers because backscattering is the communication technique usually used by RFID systems operating at UHF and microwave frequencies, whereas inductive coupling is used by RFID systems operating at LF or HF.

C is incorrect because scattering is not a communication technique; it's a characteristic that affects the propagation of wave.

D is incorrect because link margin is not a communication technique; it's a physical quantity that measures the ratio of effective signal strength received to minimum signal strength received.

7. Answer: B

B is the correct answer because an RF signal (or any EM wave, for that matter) travels the fastest through free space (with velocity of light through a vacuum).

A, C, and **D** are incorrect answers because EM waves travel more slowly in a medium than they do in the free space.

8. Answer: D

D is the correct answer because impedance mismatch does not create diffraction.

A, B, and **C** are incorrect answers because an impedance mismatch between the antenna and the transmission line will reflect the EM wave, which will interfere with the incoming EM wave and generate a standing wave.

9. Answer: B

B is the correct answer because the electric field vector and the magnetic field vector of an EM wave are perpendicular to the direction of the wave propagation.

A is an incorrect answer because it's a true statement: A horizontally polarized wave travels parallel to the surface of the Earth, whereas a vertically polarized wave travels perpendicular to the surface of the Earth.

C is an incorrect answer because the electric field vector and the magnetic field vector of an EM wave are perpendicular to the direction of the wave propagation.

D is incorrect because it's a true statement: Horizontal polarization and vertical polarization are two cases of linear polarization.

Chapter 3

1. Answer: A and B

A and **B** are the correct answers because the standard antenna for transmitting low-frequency waves would need to be impractically large. Therefore, the tags at low frequencies (LF and HF) use inductive coupling that allows them to use an inductive coil to transfer energy.

C and **D** are incorrect because UHF and microwave frequencies are generally used to achieve large read ranges, which will not be allowed by inductive coupling.

2. Answer: A and B

A and **B** are the correct answers because passive tags do not have batteries.

C, D, E, and **F** are incorrect answers because active tags and semipassive tags have their own batteries.

3. Answer: B and E

B and **E** are the correct answers because water can absorb UHF waves and metals can reflect the waves.

A, C, and **D** are incorrect because paper and plastic are relatively safer for EM wave propagation.

4. Answer: B

B is the correct answer because shadowing refers to the situation that occurs when a tag cannot be read because the tagged item is behind another item.

A is incorrect because a read by a reader outside its intended zone is called a ghost read or a phantom read.

C and b are incorrect answers because these are the wrong terms for this effect.

5. Answer:

A ■ 5
B ■ 1, 2, 3
C ■ 4

6. Answer: D

D is the correct answer because an insert is an inlay inserted between a label in the front and an adhesive layer in the back.

A, B, and **C** are incorrect because these do not closely describe an insert.

7. Answer: A

A is the correct answer because an LF passive tag has a read range of about 50 cm.

B, C, and **D** are incorrect because these ranges are too high for a low-frequency passive tag.

8. Answer: C

C is the correct answer because LF waves, due to their high wavelengths, can penetrate otherwise absorptive materials.

A, B, and **D** are incorrect because high-frequency waves, due to their shorter wavelengths, will be absorbed by the body fluids.

9. Answer: A

A is the correct answer because active tags, due to their size and battery, can have larger memory.

B, C, and **D** are incorrect because passive and semipassive tags usually have less memory than active tags.

Chapter 4

1. Answer: B

B is the correct answer because the *KILL* command disables a tag from communicating back to the reader.

A, C, and **D** are incorrect answers because these are the wrong descriptions of the *KILL* command.

2. Answer: D

D is the correct answer because an interrogator is not attached to an item; it is used to read the tags that are attached to items.

A, B, and **C** are incorrect answers because communicating with a tag is the basic functionality of an interrogator, but some interrogators can also communicate with other interrogators and with the host computers.

3. Answer: B

B is the correct answer because the serial connection setup with a cable usually uses the RS232 standard for communication.

A and **C** are incorrect because network connections (through cable or wireless) do not use RS232.

D is incorrect because a reader either has a serial connection or a network connection with the host computer.

4. Answer: A

A is the correct answer because shadowing refers to one tagged item hiding (shadowing) another tagged item, thereby preventing the reader from reading it.

B, C, and **D** are incorrect answers because these are the wrong descriptions of shadowing.

5. Answer: D

D is the correct answer because locking the tags is not an effect of dense environments. Tags are locked by issuing the *LOCK* command from the readers.

A, B, and **C** are incorrect answers because reader collision is an effect of a dense interrogator environment, and tag collisions and shadowing are the effects of a dense tag environment.

6. Answer: B

B is the correct answer because the tag read rate of an interrogator cannot be adjusted directly by configuring the reader since it depends on many external factors, such as the tag travel speed.

A is incorrect because you can use tag population management commands such as *SELECT* to tell the reader to read only certain kinds of tags.

C and **D** are incorrect because you can select and configure the protocols and you can adjust the output power to some extent.

7 Answer:

A ▪ 2

B ▪ 1

C ▪ 4

D ▪ 3

8. Answer: C

C is the correct answer because by configuring the interrogator to search only for a certain type of tag, you improve the read cycle rate and therefore the read rate.

A is incorrect because for the given situation, it's not the correct solution.

B is an incorrect answer because this will effectively set the read rate to zero; no tag entering the zone (class 0 tag) will be read.

D is an incorrect answer because searching for all tag classes will slow the read cycle and therefore the read rate.

9. Answer: A

A is the correct answer because reader collisions occur in a dense reader environment and can result in signal interference and multiple reads of the same tag.

B is incorrect because shadowing only blocks a tag from being read.

C is an incorrect answer because low power emission by itself may hamper the read efficiency but is not the most likely cause for the given problem.

D is an incorrect answer because this is not the most likely cause of the given problem.

10. Answer: B

B is the correct answer because using the same protocol is not the reason for overlapping interrogation zones, and using different protocols is not going to solve this problem.

A, C, and **D** are incorrect because all these methods will possibly help solve the problem of interrogation zone overlap.

Chapter 5

1. Answer: A

A is the correct answer because Europe is included in regulatory region 1.

B and **C** are incorrect because Europe is included in regulatory region 2.

D is incorrect because there are only three regulatory regions for RFID.

2. Answer: C

C is the correct answer because 125–134 is the most commonly used frequency range for RFID devices in the LF band.

A and **D** are incorrect because these frequencies are not in the LF range.

B is incorrect because 200 KHz, although in the LF range, is not a commonly used frequency for the RFID systems.

3. Answer: A

A is the correct answer because if the power is too high, the RF waves from the RFID device will travel farther and will interfere with the waves from other devices in the vicinity.

B and **C** are incorrect answers because saving energy and helping the environment are not the motivations behind regulating the antenna power emission.

D is incorrect because it is the RFID operating frequency that is regulated to avoid disrupting the existing RFID services.

4. Answer: D

D is the correct answer because air interface protocols deal with data communication and not with data storage or formatting.

A, B, and **C** are incorrect answers because all these are aspects of data communication between an interrogator and a tag.

5. Answer: A

A is the correct answer because air interface protocols deal with the communication between interrogators and tags.

B is an incorrect answer because tag data formats standards deal with formatting the data on the tags.

C is an incorrect answer because RS232 is used for serial communication between an interrogator and host computer.

D is an incorrect answer because TCP/IP is a suite of protocols on which the Internet is based.

6. Answer: C

C is the correct answer because EPCglobal was launched to standardize and commercialize the EPC technology developed by the Auto-ID Center at MIT.

A is an incorrect answer because ITU is a regulatory body that organized the world into three regulatory regions.

B is an incorrect answer because ISO develops standards for RFID and several other industries.

D is an incorrect answer because the FCC regulates RFID in the United States.

7. Answer: B

B is the correct answer because the EPC number does not contain the CRC field.

A, C, and **D** are incorrect answers because an EPC number contains a header, manager number, object class, and serial number.

8. Answer: D

D is the correct answer because passive tags use the power from the radiated signal from the reader as their operating power.

A and **C** are incorrect because the passive tag is the most affected tag by the power regulations.

B is incorrect because there is no tag type called an LF tag.

9. Answer: B

B is the correct answer because Dynamic Host Configuration Protocol (DHCP) is part of the TCP/IP (Internet) protocols, not part of the EPCglobal network, even though the EPCglobal network can use the Internet.

A, C, and **D** are incorrect because EPC, an ID system for tags and readers, and object name service are all parts of the EPCglobal network.

10. Answer: B

B is the correct answer because

$SAR = C \times E^2/d$

A, C, and **D** are incorrect answers because these are the wrong statements about SAR.

Chapter 6

1. Answer: A

A is the correct answer because an isotropic antenna is a perfectly omnidirectional antenna, which is a theoretical reference used for calculating quantities such as ERP.

B, C, and **D** are incorrect answers because these antennas are not perfectly omnidirectional.

2. Answer: B

B is the correct answer because the footprint of an antenna is a three-dimensional ground area over which the antenna delivers a specified amount of signal power under specified conditions.

A, C, and **D** are incorrect because these are the wrong terms for this description.

3. Answer: B

B is the correct answer because the read robustness or efficiency refers to the number and success of reads.

A is incorrect because read range is one of many factors that contribute to read efficiency.

C is incorrect because failed read attempts do not increase read efficiency.

D is incorrect because read efficiency does not refer to the signal strength.

4. Answer: C

C is the correct answer because a reader needs to support Simple Network Management Protocol (SNMP) if it is to be controlled and managed remotely.

A is incorrect because there is no standard protocol named RemoteMan used by readers.

B is incorrect because Simple Mail Transfer Protocol (SMTP) is the protocol used to transfer e-mails.

D is incorrect because Hypertext Transfer Protocol (HTTP) is the protocol on which the World Wide Web is based.

5. Answer: D

D is the correct answer because a coaxial cable can be useful for frequencies as high as 3 GHz.

A, B, and **C** are incorrect answers because a coaxial cable can be used as a transmission line for operating frequencies as high as 3 GHz.

6. Answer:

A ▪ 4
B ▪ 3
C ▪ 2
D ▪ 1

7. Answer: A

A is the correct answer because the UHF signals, due to their short wavelengths, are easily absorbed by materials such as water.

B, C, and **D** are incorrect because these are the advantages of the high-frequency signals.

8. Answer: C

C is the correct answer because to facilitate real-time processing, the RTLS requires active tags.

A, B, and **D** are incorrect answers because for these applications, passive UFH tags will work just fine since the UFH will provide good enough read range.

9. Answer: A

A is the correct answer because LF systems, due to their larger wavelengths, can work well around water.

B, C, and **D** are incorrect answers because higher-frequency signals, due to their shorter wavelengths, can be easily absorbed by water.

Chapter 7

1. Answer:

A ▪ 3
B ▪ 1
C ▪ 4
D ▪ 2

2. Answer: C

C is the correct answer because you can determine the interrogation zone by taking signal strength measurements with a spectrum analyzer around the antenna.

A, B, and **D** are incorrect answers because a site blueprint is a good start for a site analysis, but it's not enough to determine the interrogation zones.

3. Answer: B

B is the correct answer because the device that you need for site analysis is called a *spectrum analyzer*.

A is an incorrect answer because you need a cart to move your testing equipment around the site.

C is an incorrect answer because the site blueprint helps you visualize the site infrastructure.

D is an incorrect answer because a portable computer is used to record the collected data.

4. Answer: B

B is the correct answer because the site blueprint will help you visualize the site infrastructure.

A is an incorrect answer because a spectrum analyzer is used when you are performing a site survey.

C is an incorrect answer because a site map will not show you the site infrastructure.

D is incorrect because interrogation zones are tested and designed during site survey and analysis, and they cannot show you the whole site infrastructure anyway.

5. Answer: C

C is the correct answer because the full Faraday cycle analysis (FFCA) is a process to collect data regarding the EM waves in a site environment over a full business cycle, typically 24 to 48 hours, which will include all the normal operations involving RF bands about which the data will be taken.

A, B, and **D** are incorrect answers because these are the wrong names for the said process.

6. Answer: D

D is the correct answer because creating an air gap between a tag and the metallic surface will increase the read range of the tag for a given frequency and power emission.

A, B, and **C** are incorrect answers because you do not change the operating frequency of an installed system just to accommodate some tags. Besides, operating frequencies and maximum power emissions are regulated.

7. Answer: A

A is the correct answer because usually near the dock doors the signal coverage area will be maximum due to the absence of sources of adverse effects (absorption, reflection, and interference) such as metallic equipment and the like.

B and **D** are incorrect answers because metal will reflect the signal.

C is an incorrect answer because liquids will absorb the energy from the RF signals.

8. Answer: C

C is the correct answer because cost analysis is not part of site analysis, although cost could be a factor to consider in the system design.

A, B, and **D** are incorrect answers because identifying sources of interference and marking interrogation zones (coverage areas) are essential parts of site analysis. The system design is not final before and without the system analysis.

9. Answer: B

B is the correct answer because increasing the antenna size is not a solution to interference.

A, C, and **D** are incorrect answers because all these are possible solutions to interference.

Chapter 8

1. Answer: A

A is the correct answer because you can take a handheld interrogator to the tagged objects and read them from varied distance and angles (of course, within the read range).

B is an incorrect answer because although you can move the vehicle, the interrogator is still fixed on the vehicle. So, you do not have as much flexibility as in the case of a handheld interrogator.

C is an incorrect answer because the readers are fixed on the door.

D is an incorrect answer because there is no such thing as a remote portal.

2. Answer:

A ▪ 4
B ▪ 1
C ▪ 2
D ▪ 3

3. Answer: A

A is the correct answer because the National Electrical Manufacturers Association (NEMA) provides standards and enclosures for electrical equipment.

B is an incorrect answer because EPCglobal develops standards for the EPCglobal network.

C is an incorrect answer because the Federal Communications Commission (FCC) regulates RFID in the United States.

D is an incorrect answer because the Standardization Administration of China (SAC) issues regulations for RFID in China.

4. Answer: C

C is the correct answer because a spectrum analyzer is used to make measurements in the interrogation zone.

A and **B** are incorrect answers because RFID encoders and oscilloscopes are not used to measure the RF loss contour.

D is incorrect because there is no such device as a contour mapper.

5. Answer: A

A is the correct answer because an uninterruptible power supply is a device that maintains a continuous supply of electric power to the connected equipment by supplying power from a battery when utility power becomes unavailable.

B, C, and **D** are incorrect answers because they do not refer to UPS.

6. Answer: A

A is the correct answer because a ground loop creates a loop with current in **it and does not act as a sink for the electrostatic charge.**

B, C, and **D** are incorrect answers because all these are valid methods to protect against ESD.

7. Answer: C

C is the correct answer because a passive tag does not have its own power source.

A, B, and **D** are incorrect answers because all these devices can use a battery as a power source.

8. Answer: D

D is the correct answer because the equipment must be tested before and after installation.

A, B, and **C** are incorrect answers because both pre-install and post-install tests must be performed on the equipment.

9. Answer: B

B is the correct answer because if you conform to the safety regulations, there should not be too much radiation.

A, C, and **D** are incorrect answers because all these are valid steps to take in implementing safety.

Chapter 9

1. Answer: C and D

C and **D** are the correct answers because a light tree and a horn are output devices.

A and **B** are incorrect answers because a motion sensor and a photo eye are input devices.

2. Answer: A

A is the correct answer because in this situation we need a beacon, an active tag that is emitting signal continuously. When it passes through an interrogation zone, the reader will start reading after getting a signal from it.

B and **C** are incorrect answers because passive and semipassive tags cannot emit signals on their own initiative.

D is an incorrect answer because you do need a tag on an object that needs to be tracked.

3. Answer: C

C is the correct answer because a photo eye and motion sensors that act as input devices to an RFID system are sensors.

A, B, and **D** are incorrect answers because readers, tags, and host computers are not ancillary devices.

4. Answer: B

B is the correct answer because the printer is connected to a port on a computer.

A and **C** are incorrect answers because a printer is not connected to a reader or a tag. It has an encoder as a part of it, and it prints the smart labels that contain tags.

D is an incorrect answer because the EPCglobal network is not a physical device or a LAN.

5. Answer: D

D is the correct answer because direct thermal is a printing technique that produces a printed image using heat-sensitive paper when the paper passes over the thermal print head.

A is an incorrect answer because tamp-down is a label-placing technique in which the label is simply pressed against the item.

B is an incorrect answer because wipe-on is a label-placing technique in which the label is temped down with the help of a foam roller.

C is an incorrect answer because blow-on is a label-placing technique in which the label is blown onto the item by a blast of air.

6. Answer:

A ▪ 3

B ▪ 1

C ▪ 2

D ▪ 4

E ▪ 5

Chapter 10

1. Answer: D

D is the correct answer because you might need to perform root cause analysis iteratively to prevent the problem from recurring.

A, B, and **C** are incorrect answers because these are correct statements about root cause analysis.

2. Answer: C

C is the correct answer because it's perfectly fine to make multiple read attempts to ensure that the tag is read.

A, B, and **D** are incorrect answers because these are valid reasons for a tag failure.

3. Answer: B

B is the correct answer because you do not identify improperly placed tags by varying the power emitted by the reader.

A, C, and **D** are incorrect answers because these are valid methods for identifying improperly placed tags.

4. Answer: D

D is the correct answer because the *ping* command is used to test the reachability of any device connected to the network.

A is an incorrect answer because *ipconfig* will give information about the network connection of the local host.

B and **C** are incorrect answers because these are not standard network commands.

5. Answer: B

B is the correct answer because intrusive monitoring might be needed to get a deeper insight into the system to find the root cause of a problem.

A and **C** are incorrect answers because nonintrusive and status monitoring may not be enough to find the root cause of a problem.

D is an incorrect answer because performance monitoring may be needed to find the root cause of a problem.

6. Answer: C

C is the correct answer because *ping* is a TCP/IP command.

A, B, and **D** are incorrect answers because *ping* is a TCP/IP command.

7. Answer: C

C is the correct answer because *ping* is a TCP/IP command.

A, B, and **D** are incorrect answers because *ping* is a TCP/IP command.

8. Answer: B

B is the correct answer because high power emission by an antenna is not the most likely reason for read errors.

A, C, and **D** are incorrect answers because all these may be indicated by read errors. For example, low signal strength will provide shorter read range, and improper placement of antennas and tags will prevent the reader from reading the tags.

9. Answer: A and C

A and **C** are the correct answers because the change in read error rate and smaller values of mean time between two failures indicate instability of the system.

B and **D** are incorrect answers because average tag traffic volume and read error rates by themselves are not any indication of stability or instability of a system.

E is an incorrect answer because each RFID system runs at a predetermined operating frequency and it does not measure the stability of the system.

10. Answer: B

B the correct answers because monitoring is hardly a reason for hardware failure.

A, C, and **D** are incorrect answers because all these are possible reasons for hardware failures.

Appendix B

Questions

1. Which formula will you use to calculate the wavelength of an RF wave whose frequency is f and that is traveling through free space?
   A. $\lambda = f/c$
   B. $\lambda = fc$
   C. $\lambda = c/2f$
   D. $\lambda = c/f$

2. RFID systems typically use the transmission line with the following impedance:
   A. 10 Ω
   B. 50 Ω
   C. 100 Ω
   D. 70 Ω

3. Which characteristic of an RF wave affects the performance of an RFID system in the vicinity of metal cans?
   A. Reflection
   B. Refraction
   C. Diffraction
   D. Absorption

4. For signal strength V_s and noise strength V_n, the signal to noise ratio, SNR, can be calculated in decibels by using the following formula:
   A. $SNR = 20 \log (V_s/V_n)$
   B. $SNR = 10 \log (V_s/V_n)$
   C. $SNR = V_s/V_n$
   D. $\log (V_s/V_n)$

5. Which of the following is not the characteristic that affects the propagation of an RF wave through imperfect free space?
   A. Reflection
   B. Absorption
   C. Diffraction
   D. ERP

6. Which of the following antenna types are typically not used in RFID systems?
   A. Isotropic
   B. Directional
   C. Dipole
   D. Monostatic
7. Frequency shift keying is an example of which of the following? (Choose two.)
   A. Digital modulation
   B. Amplitude modulation
   C. Phase modulation
   D. Frequency modulation
8. Which of the following generates a standing wave?
   A. Impedance mismatch between antenna and transmission line
   B. Water or products with water content
   C. Metals
   D. Objects with sharp edges
9. In an environment of poor directivity, a reader might end up reading a tag outside its zone. This is called:
   A. A ghost read
   B. The shadowing effect
   C. High efficiency
   D. An extra read
10. A UHF passive tag will typically use which of the following for communication?
    A. Inductive coupling
    B. Backscattering
    C. Reflective coupling
    D. Passive coupling
11. Which of the following is the correct read range of a UHF passive tag?
    A. Less than a few centimeters
    B. Less than one foot
    C. Less than a meter
    D. Up to a few meters
12. The design of a tag size is typically limited by which of the following?
    A. The memory size
    B. The antenna size
    C. The chip size
    D. The substrate size
13. Which of the following RF waves will be most vulnerable to absorption by material such as water?
    A. LF
    B. HF
    C. UHF
    D. Microwave
14. Which of the following tags can offer sensor functionality?
    A. Class 0
    B. Class 1
    C. Class 2
    D. Class 3

15. Match the items in the second column of the table to the items in the first column.

Material	Effect on RF
A. Corrugated cardboard	1. Absorption
B. Glass	2. Attenuation
C. Liquid	3. Detuning due to dielectric effect
D. Group of cans	4. Reflection and multiple paths propagation
E. Animal	
F. Plastic	

16. With increase in frequency, the read range of a tag:
    A. Decreases
    B. Increases
    C. Remains unaffected
    D. Increases only if it's an active tag
17. A UFH passive tag is unable to read the boxes in the center of the pallet. What is the most likely cause of this problem?
    A. Shadowing
    B. Passive backscattering
    C. Inductive coupling
    D. Impedance mismatch
18. A reader's antenna is generating horizontally polarized radio waves. It should be able to communicate with tags that have antennas with what kind of orientation? (Choose two.)
    A. Parallel to the propagation direction of the incoming wave
    B. Vertical to the surface of the Earth
    C. Parallel to the incoming field
    D. Parallel to the surface of the Earth
19. Which of the following dipole antennas consists of two or more straight conductors that are connected in parallel?
    A. Dual dipole antenna
    B. Folded dipole antenna
    C. Dipole antenna
    D. Quarter wavelength dipole antenna
20. If the tags are not properly aligned, they will not get much energy from the waves emitted by which type of reader antenna?
    A. Circularly polarized antenna
    B. Helical antenna
    C. Linearly polarized antenna
    D. Inductive coupling antenna
21. Which of the following tag types will have the largest read range?
    A. Class 0
    B. Class 1
    C. Class 2
    D. Class 4

22. What is the minimum number of antennas that most readers support?
    A. 1
    B. 2
    C. 3
    D. 4
23. You want to make sure that maximum transfer of power occurs between the source and the antenna through the transmission line. Which of the following will help you do that?
    A. Ensure the match between the input impedance of the antenna and characteristic impedance of the transmission line.
    B. Ensure there is no water or mud in the vicinity of the antenna.
    C. Use coaxial cable as a transmission line.
    D. Use shielded pair cable as a transmission line.
24. Match the items in the second column of the table to the items in the first column.

Mounting Equipment (Portal)	Function
A. Conveyor	1. Cost saving by eliminating the need of cash counter operators
B. Dock door	2. Provides the reader with multiple opportunities to read the tags in different orientations
C. Stretch wrap station	3. Pallet-level tracking; works in the presence of other electronic devices
D. Point of sale	4. Case-level tracking; works better with multiple antennas

25. You are designing a mount portal that will work under conditions of mud and snow. Which of the following operating frequencies are most suitable for your RFID system under these conditions? (Choose two.)
    A. 125 KHz
    B. 13.56 MHz
    C. 433 MHz
    D. 2.44 GHz
    E. 5.8 GHz
    F. 915 MHz
26. Tag ruggedness is needed for which of the following reasons?
    A. To avoid reflection
    B. To increase the read range
    C. To withstand the harsh environmental conditions
    D. To avoid impedance mismatch
27. All of the following are factors in selecting the operating frequency for RFID systems *except*:
    A. Impedance mismatch
    B. Read range
    C. Application type
    D. Operating conditions
28. Europe is included in which of the following regulatory regions for RFID?
    A. Region I
    B. Region II
    C. Region III
    D. Region IV

29. What is the most commonly used RFID frequency in the HF range?
    A. 13.56 MHz
    B. 200 KHz
    C. 125–134 KHz
    D. 13.56 KHz
30. What is the purpose of regulating the power emitted by the antenna of an RFID device?
    A. To avoid disrupting existing RF services
    B. To help keep the radiation level within safe limits
    C. To save energy
    D. To help the environment
31. Air interface protocols define rules for all of the following *except*:
    A. Communication between a reader and a tag
    B. Anticollision algorithms
    C. Commands for reading from and writing to tags
    D. Electronic product code
32. What kind of standards format an EPC number?
    A. Air interface protocols
    B. Tag data format standards
    C. RS232
    D. TCP/IP
33. Which organization regulates RFID technology in the United States?
    A. ITU
    B. ISO
    C. EPCglobal
    D. FCC
34. An EPC number typically contains all of the following fields except:
    A. Manager number
    B. Header
    C. Price
    D. Object class
35. Which of the following tag classes is most affected by regulating the maximum radiated power?
    A. Class 2
    B. Class 3
    C. Class 4
    D. Class 5
36. All of the following are the services promised by the EPCglobal network *except*:
    A. Identification
    B. Lookup
    C. Querying
    D. Domain name system
37. Specific absorption rate (SAR) is:
    A. Linearly proportional to the conductivity of the body tissue
    B. Directly proportional to the square of the conductivity of the body tissue
    C. Inversely proportional to the conductivity of the body tissue
    D. Inversely proportional to the square of the conductivity of the body tissue

38. The *LOCK* command on an interrogator is used to:
    A. Lock the data on the tag so that no reader can read it
    B. Lock the reader so that nobody can issue a command on it
    C. Disable writing to a tag
    D. Disable reading from a tag
39. All of the following are functions of an interrogator *except*:
    A. Provide power to a passive tag
    B. Provide power to an active tag
    C. Change the information stored on a tag
    D. Encode the information into an outgoing RF wave and decode the information from an incoming RF wave
40. While talking about connecting a reader to the host computer, your manager mentions connecting the reader directly to the host computer with a cable. What kind of connection is he referring to?
    A. Serial connection
    B. Network connection
    C. Wireless connection
    D. Parallel connection
41. Which of the following is not true about shadowing?
    A. It is a situation created by dense tag environment in which a tagged item blocks the reader signal from reaching another tagged item hiding behind it.
    B. Shadowing can occur when a tag acts as a reflector and thereby reflects (or blocks) an RF signal from reaching another tagged item.
    C. Shadowing can cause interference.
    D. It refers to the situation that occurs when a tag is too close to an antenna to be read properly.
42. Which of the following is not an effect of a dense interrogator environment?
    A. Reader collision
    B. Shadowing
    C. Signal interference
    D. Multiple reads of the same tag by multiple readers
43. All of the following are the configuration settings of an interrogator *except*:
    A. Change the reader class
    B. Event notification
    C. Host management
    D. Rename the interrogator
44. Match the items in the second column of the table to the items in the first column.

RFID Command	Action
A. INVENTORY	1. Group the tags and select a group of tags for the interrogator.
B. SELECT	2. Initiate communication with a tag.
C. QUERY	3. Get access to an individual tag in a group.
D. ACCESS	4. Identify an individual tag in a group.

45. You know that an interrogator zone at your site always gets items in groups of 10 at a time. You have noticed that the read rate of this zone is not that great. What can you do to increase the read rate?
    A. Increase the power of the interrogator to maximum.
    B. Configure the interrogator to query tags in groups of 10.
    C. Set the interrogator to search for class 0 tags only.
    D. Set the interrogator to use only one antenna.
46. You have set up an interrogator in a small warehouse. You have tested it at various places in the warehouse. You have observed that at some places it does not read all the tags in its interrogation zone. What do you think the most likely problem is?
    A. Reader collision
    B. Shadowing
    C. Low power emission by the reader
    D. Interrogation zone overlap
47. How does the read range of a handheld interrogator usually compare to that of a vehicle-mount interrogator?
    A. Less than the read range of a vehicle-mount interrogator
    B. The same as the read range of a vehicle-mount interrogator
    C. Greater than the read range of a vehicle-mount interrogator
    D. The same or greater than the read range of a vehicle-mount interrogator
48. Match the items in the second column of the table to the items in the first column.

Item	Description
A. Ambient EM noise	1. Helps to visualize the infrastructure of the site
B. Site blueprint	2. Identifies the sources of interference by measuring the signal strength
C. Spurious emission	3. Electromagnetic noise existing at a specific location
D. Spectrum analyzer	4. Interfering radiation transmitted outside the operating frequency band in the form of narrowband signals or wideband noise

49. Which of the following is used to measure RF output from circuits, devices, and instruments?
    A. CAD drawing
    B. Site blueprint
    C. Spectroscope
    D. Spectrum analyzer
50. While planning for site analysis, you need all of the following *except*:
    A. Two stands
    B. Spectrum analyzer
    C. Site blueprint
    D. Portable computer
    E. AEN
51. When will you deal with the site blueprint?
    A. Before the site analysis
    B. After the site analysis
    C. During the site analysis
    D. Before, after, and during the site analysis

52. Full Faraday cycle analysis (FFCA) is:
    A. Just another name for site analysis
    B. A process to collect data regarding the EM waves in a site environment over a full business cycle, which is typically 24 to 48 hours
    C. A measurement unit for RF interference
    D. Used only to analyze a site that sells electronic devices
53. You are implementing an RFID system in a shipping area. Some of the items that you are going to tag are metallic boxes—that is, you are going to place tags on metallic surfaces. Which of the following solutions can you implement to optimize read performance?
    A. Increase the power of the tags.
    B. Change the operating frequency of the readers.
    C. Use nonmetallic spacers between the tag and the metal.
    D. Increase the power of the readers.
54. You are looking at the blueprint of a warehouse to visualize the site infrastructure before a site survey. Which of the following areas will have maximum signal coverage?
    A. Racks with metal shelves
    B. Dock doors
    C. A room with cleaning liquids
    D. Trolleys
55. All of the following are the results of site analysis *except*:
    A. Blueprints
    B. System deployment
    C. Reports
    D. Marked interrogation zones
56. All of the following site analysis results should go into the blueprint *except*:
    A. Maximum signal coverage area
    B. Minimum signal coverage area
    C. Location for metallic equipment
    D. Estimated cost for installing readers in different places
57. In which of the following RFID portal solutions is tag speed an important factor that affects the number of tags read by a reader?
    A. Smart shelves
    B. Vehicle-mount interrogator portal
    C. Dock door portal
    D. Conveyor portal
58. Match the items in the second column of the table to the items in the first column.

RFID Portal	Most important function or consideration
A. Conveyor	1. Tag orientation
B. Dock door	2. Mechanical shocks and vibrations
C. Forklift	3. Case-level tracking
D. Shelf	4. Pallet-level tracking

59. What is NEMA?
    A. An organization that develops standards for the EPCglobal network
    B. An organization that regulates RFID in the United States
    C. An organization that provides standards for electrical equipment and enclosures for electrical equipment
    D. An organization that regulates RFID in China

60. RF path loss contour mapping (PLCM) helps accomplish all of the following *except*:
    A. Determining where the antenna should be installed in an interrogation zone
    B. Determining the boundaries of an interrogation zone
    C. Configuring and fine-tuning a reader
    D. Determining which communication protocol should be used by the reader

61. Which one of the following power supplies transmits electrical power, along with the data to remote devices over standard twisted-pair cable in an Ethernet network?
    A. UPS
    B. Linear power supply
    C. Switched mode power supply
    D. POE

62. How will you avoid ground loops?
    A. By making sure that the RFID devices are not grounded
    B. By using a short conductor for grounding
    C. By making sure that each device is grounded to the separate points on the ground
    D. By using UPS

63. Which of the following is not the way to protect devices or personnel from ESD?
    A. Air ionization
    B. Keeping relative humidity as high as possible
    C. Plastic handles on the tools
    D. Conductive mats

64. A ground loop occurs when:
    A. There are too many electronic devices in the same room
    B. Multiple devices are connected to the same grounding system
    C. There are multiple ground connections between two devices
    D. The conductor used to ground a device is too short

65. You are installing an RFID system. Which of the following is not a necessary consideration in ensuring safety?
    A. ESD
    B. Ground loops
    C. Number of readers
    D. Safety regulations

66. Which *two* of the following are ancillary input devices for an RFID system?
    A. Motion sensor
    B. Photo eye
    C. Light tree
    D. Horn

67. Which of the following is not an essential component of an RTLS that tracks moving objects?
    A. Stationary readers
    B. Moving tags
    C. Monitoring system
    D. Smart label printer

68. Which of the following is an ancillary device that is typically used on conveyors to trigger reads when the load is arriving and is also used to verify whether the tags are placed properly?
    A. Encoder
    B. Photo eye
    C. Motion sensor
    D. Light tree

69. Which printer property lets you configure the position of the smart label on the printer?
    A. Media handling
    B. Label positioning
    C. Orientation
    D. Paper feed shift

70. Which of the following printing techniques requires ribbon?
    A. Direct thermal
    B. Thermal transfer
    C. Wipe-on
    D. Temp-down

71. The label-placing technique in which a label is placed on an item without direct contact with the item is called:
    A. Tamp-down
    B. Wipe-on
    C. Blow-on
    D. Direct thermal

72. Match the items in the second column of the table to the items in the first column.

Device	Function
A. RFID printer	1. Provides capability to program smart labels
B. Label applicator	2. In conjunction with the software, can help determine bad tags or badly placed tags
C. Photo eye	3. Prints a barcode on a smart label and writes data on to the tag inside the smart label
D. RFID printer encoder	4. Attaches labels to items that need to be tracked

73. Which of the following is not true about root-cause analysis (RCA)?
    A. Applying corrective measures to root causes of a problem is more effective than simply addressing the symptoms of the problem.
    B. To get optimal results, RCA must be performed in a systematic fashion, and its conclusions must be backed up by evidence.
    C. There is one root cause for a given problem.
    D. The goal of root-cause analysis is to prevent a problem from recurring.

74. All of the following are reasons for RFID tag failures *except*:
    A. Harsh environmental conditions
    B. Multiple tags on an item
    C. Improperly placed tag on an item
    D. Dense tag environment

75. Which of the following is not an element of the EPCglobal network?

A. EPC
B. EPC tags
C. Object Naming Service (ONS)
D. Dynamic Host Configuration Protocol (DHCP)

76. A reader is not working properly. You are thinking of troubleshooting it. What is the first step that you will take?
    A. Replace the reader and send it for repair.
    B. Restart the reader.
    C. Shut down the whole RFID system of which the reader is a part.
    D. Restart all the components of the RFID system.

77. Which of the following is not a type of monitoring?
    A. Status monitoring
    B. Performance monitoring
    C. Root cause monitoring
    D. Nonintrusive monitoring

78. SNMP is:
    A. A TCP/IP protocol used to manage devices connected to the TCP/IP networks
    B. An EPCglobal standard used to manufacture passive tags
    C. A TCP/IP protocol used to send and receive e-mails
    D. An RFID security standard

79. The read error rate can indicate any of the following except:
    A. Low signal strength
    B. Improper placement of tags
    C. Signal absorption
    D. Low tag traffic volume

80. Which of the following measures the stability of an RFID system?
    A. Change in read error rate
    B. Average tag traffic volume through an interrogation zone
    C. Read error rate
    D. Operating frequency

81. An RFID system can experience hardware failure due to any of the following *except*:
    A. ESD
    B. Unregulated power supply
    C. MTBF
    D. Harsh environmental conditions

Answers and Explanations

1. Answer: D

D is the correct answer because the relationship between frequency and wavelength is $f = c/\lambda$ that is, $\lambda = c/f$.

A, B, and **C** are incorrect answers because none of them correctly represents the relationship between frequency and wavelength.

2. Answer: B

B is the correct answer because RFID systems typically use a 50-Ω coaxial cable as a transmission line.

A, C, and **D** are incorrect because RFID systems typically use a 50-Ω coaxial cable as a transmission line.

3. Answer: A

A is the correct answer because metals are good reflectors.

B is an incorrect answer because refraction causes the change in direction of a wave at an interface between two dissimilar media, but the wave does not return to the medium from which it hit the interface.

C is incorrect because diffraction refers to the bending of an EM wave when it strikes sharp edges or when it passes through a narrow gap (slit).

D is incorrect because absorption will usually occur in the presence of water or food products that contain water.

4. Answer: A

A is the correct answer, and **B, C,** and **D** are incorrect answers, because SNR is defined by:

$$\mathbf{SNR = (A_s/A_n)^2}$$

By definition of dB, this can be expressed in decibels as follows:

$$\text{SNR (dB)} = 10 \log (A_s/A_n)^2 = 20 \log (A_s/A_n)$$

5. Answer: D

D is the correct answer because ERP is the characteristic of an antenna.

A, B, and **C** are incorrect answers because all these characteristics affect the propagation of RF waves through free space.

6. Answer: A

A is the correct answer because the communication in an RFID system is directed and not broadcast.

B, C, and **D** are incorrect answers because these antennas can be used in RFID systems, since they are directional antennas.

7. Answer: A and D

A and **D** are the correct answers because frequency shift keying (FSK) is the frequency modulation applied to a digital signal.

B and **C** are incorrect because FSK is neither amplitude modulation nor phase modulation.

8. Answer: A

A is the correct answer because an impedance mismatch between the antenna and the transmission line will reflect the EM wave, which will interfere with the incoming EM wave and will generate a standing wave.

B is incorrect because water can absorb high-frequency waves.

C is incorrect because metals reflect EM waves.

D is incorrect because objects with sharp edges will cause diffraction.

9. Answer: A

A is the correct answer because a read by a reader outside its intended zone is called a ghost read or a phantom read.

B is incorrect because shadowing effect refers to the situation that occurs when a tag cannot be read because the tagged item is behind another item.

C and D are incorrect answers because these are the wrong terms for this effect.

10. Answer: B

B is the correct answer because a UHF passive tag uses backscattering.

A is incorrect because inductive coupling is used at LF and HF.

C and **D** are incorrect because there are no communication techniques named *reflective coupling* and *passive coupling*.

11. Answer: D

D is the correct answer because a UFH passive tag may have a read range of up to a few meters.

A, B, and **C** are incorrect answers because although a passive tag has shorter read range than an active tag, the read range also depends on the operating frequency.

12. Answer: B

B is the correct answer because the limit on the minimum size is usually driven by the antenna size, which in turn depends on the operating frequency.

A and **C** are incorrect because very small chip (and memory) sizes are available.

D is incorrect because substrate is used to house the chip and antenna.

13. Answer: D

D is the correct answer because the higher the frequency, the shorter the wavelength and therefore the higher the probability for absorption.

A, B, and **C** are incorrect answers because LF, HF, and UHF waves have lower frequency and therefore higher wavelength than microwaves.

14. Answer: D

D is the correct answer because class 3 tags are semipassive, have their own batteries, and can offer sensor functionality.

A, B, and **C** are incorrect answers because all these tags are passive tags and do not offer sensor functionality.

15. Answer:

A. 1

B. 2

C. 1

D. 4

E. 1

F. 3

16. Answer: B

B is the correct answer because the read range increases with an increase in frequency for both active and passive tags.

A, C, and **D** are incorrect answers because the read range increases with an increase in frequency for both active and passive tags.

17. Answer: A

A is the correct answer because the inner boxes are behind the outer boxes, and this effect is called *shadowing*.

B and **C** are incorrect because backscattering and inductive coupling are the communication techniques.

D is incorrect because impedance mismatch happens in the transmission line. If this was the cause here, it would affect the read of all the boxes.

18. Answer: C and D

C and **D** are the correct answers because the tag's antenna should be horizontal to the surface of the Earth to receive the energy of a horizontally polarized wave. Also, the antenna should be parallel to the incoming field but not parallel to the direction of propagation of the incoming wave.

A and **B** are incorrect because a vertically oriented antenna will not receive a horizontally polarized wave.

19. Answer: B

B is the correct answer because a folded dipole antenna consists of two or more straight electric conductors that are connected in parallel, and each electric conductor is half the wavelength corresponding to the frequency to be used.

A is incorrect because a dual dipole antenna consists of two dipoles only.

C is incorrect because a dipole antenna simply consists of a straight electric conductor.

D is incorrect because a quarter wavelength dipole antenna consists of a straight conductor that uses the reflective ground plane, which provides an image of the antenna to complete the dipole.

20. Answer: C

C is the correct answer because to the tag's orientation must be consistent with the polarization of the reader's antenna, especially in the case of linear polarization. For example, if a reader's antenna is horizontally polarized and tags are vertically oriented, the tags will not receive any power from the reader antenna.

A and **B** are incorrect because the tags in any orientation will receive some energy from a circularly polarized reader antenna; a helical antenna is a circularly polarized antenna.

D is incorrect because inductive coupling does not use antennas.

21. Answer: D

D is the correct answer because class 4 tags are active tags, which have greater read range than passive tags.

A, B, and **C** are incorrect answers because class 0, class 1, and class 2 tags are passive tags.

22. Answer: B

B is the correct answer because most readers support at least two antennas.

A, C, and **D** are incorrect because most readers support a minimum of two antenna23.

23. Answer: A

A is the correct answer because you must match the input impedance of the antenna with the characteristic impedance of the transmission line for maximum transfer of power. An impedance mismatch will result in a standing wave, which could decrease the power transfer.

B is incorrect because the presence of water content will affect the propagation of the wave after the antenna has transmitted it.

C and **D** are incorrect because the cable type itself might not guarantee that there will be no impedance mismatch.

24. Answer:

A. 4

B. 3

C. 2

D. 1

25. Answer: A and B

A and **B** are the correct answers because LF and HF can penetrate material such as water better than UHF and microwave frequencies can.

C, D, E, and **F** are incorrect answers because due to their smaller wavelengths, the UHF and microwave frequencies are easily absorbed by materials such as water.

26. Answer: C

C is the correct answer because ruggedness is required by a tag to withstand harsh environmental conditions such as corrosive chemicals, extremely high or low temperature, humidity, and mechanical shocks.

A, B, and **D** are incorrect because reflection, read range, and impedance mismatch are not reasons for requiring ruggedness.

27. Answer: A

A is the correct answer because impedance mismatch produces stand waves regardless of the frequency range.

B is incorrect answer because read range depends on operating frequency.

C is incorrect because often applications belonging to the same application type have the same frequency requirement.

D is incorrect because operating conditions play a role in determining the right operating frequency. For example, if the system is operating near water and mud, LF is a better solution.

28. Answer: B

B is the correct answer because the United States is included in regulatory region 2.

A and **C** are incorrect because the United States is included in regulatory region 2.

D is incorrect because there are only three regulatory regions for RFID.

29. Answer: A

A is the correct answer because 13.56 is the most commonly used frequency for RFID devices in the HF band.

B and **D** are incorrect because 200 KHz and 13.56 KHz are not in the HF range and are not commonly used frequencies for RFID systems.

C is an correct answer because 125–134 KHz is the most commonly used frequency range for RFID devices in the LF band.

30. Answer: B

B is the correct answer because the higher the power emitted by an antenna, the higher will be the specific absorption rate (SAR).

A is incorrect because it is the RFID operating frequency that is regulated to avoid disrupting the existing RFID services.

C and **D** are incorrect answers because saving energy and helping the environment are not the motivations behind regulating the antenna power emission.

31. Answer: D

D is the correct answer because air interface protocols deal with data communication, not with data storage or formatting.

A, B and **C** are incorrect answers because all these are aspects of data communication between an interrogator and a tag.

32. Answer: B

B is the correct answer because tag data formats standards deal with formatting the data on the tags.

A is an incorrect answer because air interface protocols deal with the communication between interrogators and tags.

C is an incorrect answer because RS232 is used for serial communication between an interrogator and a host computer.

D is an incorrect answer because TCP/IP is a suite of protocols on which the Internet is based.

33. Answer: D

D is the correct answer because the FCC regulates RFID in the United States.

A is an incorrect answer because the ITU is a regulatory body that organized the world into three regulatory regions.

B is an incorrect answer because the ISO develops standards for RFID and for several other industries.

C is an incorrect answer because EPCglobal was launched to standardize and commercialize the EPC technology developed by the Auto-ID Center at MIT.

34. Answer: C

C is the correct answer because the EPC number typically does not contain the price.

A, B, and **D** are incorrect answers because an EPC number contains the header, manager number, object class, and serial number.

35. Answer: A

A is the correct answer because class 2 tags are passive tags, which use the power from the radiated signal from the reader as their operating power.

B, C, and **D** are incorrect because class 3 tags are semipassive, whereas class 4 and class 5 tags are active tags, and it's the passive tags that are the most affected by the power regulations.

36. Answer: D

D is the correct answer because the domain name system (DNS) is part of the TCP/IP (Internet) protocols, not part of the EPCglobal network, even though EPCglobal network can use the Internet.

A, B, and **C** are incorrect because identification, lookup, and querying are included in the set of services offered by the EPCglobal network.

37. Answer: A

A is the correct answer because:

SAR = C x E^2/d

B, C, and **D** are incorrect answers because these are the wrong statements about SAR.

38. Answer: **C**

C is the correct answer because the *LOCK* command disables writing to a tag.

A, B, and **D** are incorrect answers because these are the wrong descriptions of the *LOCK* command.

39. Answer: B

B is the correct answer because an active tag does not use power from the interrogator; it has its own power source.

A, C, and **D** are incorrect answers because these are true statements about an interrogator. Interrogators have the functionality to power a passive tag, change the information on a tag, encode the outgoing information into a RF wave, and decode the information from an incoming signal.

40. Answer: B

A is the correct answer because the serial connection is set up by directly connecting an interrogator and a host computer with a cable.

B and **C** are incorrect because for a network connection (through a cable or wireless), you do not need to connect the devices directly with cables.

D is incorrect because a reader either has a serial connection or a network connection with the host computer.

41. Answer: D

D is the correct answer because shadowing refers to one tagged item hiding (shadowing) another tagged item, thereby preventing the reader from reading it.

A, B, and **C** are incorrect answers because these are the correct descriptions of shadowing. A tag can hide (shadow) another tag by reflecting the reader's signal and thereby blocking it from reaching another tag. The reflected signal can interfere with incident signals.

42. Answer: B

B is the correct answer because shadowing is an effect of a dense tag environment.

A, C, and **D** are incorrect answers because reader collision, signal interference, and multiple reads of the same tag are effects of a dense interrogator environment.

43. Answer: A

A is the correct answer because the tags, not the readers, have classes.

B, C, and **D** are incorrect because you can enable event notification, list the host computers to enable them to communicate with the interrogator, and rename, enable, or disable the interrogator.

44. Answer:

A. 4

B. 1

C. 2

D. 3

45. Answer: B

B is the correct answer because this will optimize the read rate by optimizing the read cycles.

A is an incorrect answer because for the given situation, it's not the correct solution.

C is an incorrect answer because you don't know whether all the tags entering the zone are class 0 tags.

D is an incorrect answer because for the given situation, this is not the correct solution.

46. Answer: B

B is the correct answer because shadowing refers to a tagged item blocking another tagged item from being read. It can happen in a dense tag environment.

A and **D** are incorrect answers because reader collisions occur in a dense reader environment and we have only one reader here.

C is an incorrect answer because the effect of low power emission should be the same throughout the warehouse, with everything else unchanged.

47. Answer: A

A is the correct answer because the handheld interrogators are typically designed for very short read ranges.

B, C, and D are incorrect because these are false statements about the read range of a handheld interrogator.

48. Answer:

A. 3

B. 1

C. 4

D. 2

49. Answer: D

D is the correct answer because a spectrum analyzer is used to measure interference and noise during the site survey.

A, B, and C are incorrect answers because site blueprints and CAD drawings are a good start for a site analysis, to help you visualize the site infrastructure, but they will not show you the measurements you need.

50. Answer: E

E is the correct answer because AEN means ambient electromagnetic noise.

A is an incorrect answer because you will need stands to support the antennas.

B is an incorrect answer because you will need a spectrum analyzer to identify interference sources.

C is an incorrect answer because a site blueprint helps you visualize the site infrastructure.

D is an incorrect answer because a portable computer is used to record the collected data.

51. Answer: D

D is the correct answer because you need a blueprint before the site analysis to visualize the site infrastructure and during and after the site analysis to enter some results into it.

A, B, and **C** are incorrect because you can use the blueprint before, after, and during the site analysis.

52. Answer: B

B is the correct answer because the full Faraday cycle analysis (FFCA) is a process to collect data regarding the EM waves in a site environment over a full business cycle, which is typically 24 to 48 hours. It includes all the normal operations involving RF bands about which the data will be taken.

A, C, and **D** are incorrect answers because these are the wrong descriptions of the FFCA.

53. Answer: C

C is the correct answer because nonmetallic spacers will create air gaps between tags and metallic boxes, and that will increase the read range of the tag for a given frequency and power emission.

A, B, and **D** are incorrect answers because you do not change the operating frequency just to accommodate some of the tags. Besides, operating frequencies and maximum power emissions are regulated and are basically selected to meet a wide range of application and environment requirements.

54. Answer: B

B is the correct answer because usually near the dock doors the signal coverage area will be at a maximum due to the absence of sources of adverse effects (absorption, reflection, and interference) such as metallic equipment.

A and **D** are incorrect answers because metal will reflect the signal.

C is an incorrect answer because liquids will absorb the energy from the RF signals.

55. Answer: B

B is the correct answer because a system deployment is not part of the site analysis; the system is installed after the site analysis.

A is an incorrect answer because you can create new blueprints or modify the existing ones to record your site analysis results.

C is an incorrect answer because the results of the site analysis should be documented in detail in a report.

D is an incorrect answer because marking interrogation zones is one of the important tasks of the site analysis.

56. Answer: D

D is the correct answer because the estimated cost for installing a reader at a specific location is usually not entered into the blueprint.

A, B, and **C** are incorrect answers because all this information can go into the blueprint.

57. Answer: D

D is the correct answer because the greater the conveyer speed, the smaller the time a tag has in the interrogation zone.

A is an incorrect answer because tags in the smart shelves are stationary.

B is an incorrect answer because in the case of a vehicle-mount portal, it's the reader that moves; the tags are stationary.

C is an incorrect answer because the speed of the tags passing through the dock doors is usually not an issue.

58. Answer:

A. 3

B. 4

C. 2

D. 1

59. Answer: C

C is the correct answer because the National Electrical Manufacturers Association (NEMA) provides standards for electrical equipment and enclosures for electrical equipment.

A is an incorrect answer because EPCglobal develops standards fro the EPCglobal network.

B is an incorrect answer because the Federal Communications Commission (FCC) regulates RFID in the United States.

D is an incorrect answer because the Standardization Administration of China (SAC) issues regulations for RFID in China.

60. Answer: D

D is the correct answer because PCLM is not used for comparing and selecting communication protocols.

A, B, and **C** are incorrect answers because PCLM helps perform all these tasks.

61. Answer: D

D is the correct answer because the Power over Ethernet (POE) technology system transmits electrical power, along with data, for example, to remote devices over standard twisted-pair cable in an Ethernet network.

A, B, and **C** are incorrect answers because these power supplies do not send power with the data.

62. Answer: B

B is the correct answer because using short enough grounding conductor will help avoid ground loops.

A is an incorrect answer because not grounding is a very unsafe option.

C is an incorrect answer because grounding devices to separate points on the ground is a perfect way to create ground loops.

D is an incorrect answer because the selection of a power supply does not help avoid a ground loop.

63. Answer: B

B is the correct answer because excessive humidity can cause problems such as corrosion, high-voltage leakage paths, and moisture contamination within the equipment.

A, C, and **D** are incorrect answers because all these are valid methods to protect against ESD.

64. Answer: C

C is the correct answer because it takes more than one connection path to create a loop between two devices.

A is an incorrect answer because the number of electronic devices in a room is, by itself, not a reason for ground loops.

B and **D** are incorrect answers because these are the ways to avoid ground loops.

65. Answer: C

C is the correct answer because the number of readers is determined by the requirement to read all the tags, not to have a dense interrogation environment. It by itself is not a safety issue.

A, B, and **D** are incorrect answers because all these are valid considerations for safety.

66. Answer: A and B

A and **B** are correct answers because a motion sensor and a photo eye are input devices.

C and **D** are incorrect answers because a light tree and a horn are output devices.

67. Answer: D

D is the correct answer because a smart label printer is not an essential component of RTLS.

A, B, and **C** are incorrect answers because you do need these three components to track moving objects.

68. Answer: B

B is the correct answer because a photo eye can detect the presence and direction of an item coming on a conveyor and trigger the reader to begin reading. Combined with software, it can also help determine a good or bad read—that is, a good or bad tag.

A, C, and **D** are incorrect answers because it's a photo eye that is typically used on the conveyors. A light tree is an output device, and an encoder is not a sensor.

69. Answer: D

D is the correct answer because paper feed shift represents the distance to advance a label or pull back when the Tear-Off Strip, Tear-Off, Peel-Off, or Cut Media Handling option is enabled.

A is an incorrect answer because media handling specifies how the printer will handle the media: peel off, tear off, cut, and so on.

B is an incorrect answer because there is typically no property called label positioning.

C is an incorrect answer because orientation specifies the image orientation that will be used when printing the label.

70. Answer: B

B is the correct answer because a printer using the thermal transfer technique uses heat to register an impression on paper via a heat-sensitive ribbon.

A is an incorrect answer because a printer using the direct thermal technique produces a printed image by using heat-sensitive paper when the paper passes over the thermal print head.

C and **D** are incorrect answers because wipe-on and tamp-down are not print techniques; they are label-placing techniques.

71. Answer: C

C is the correct answer because blow-on is a label-placing technique in which the label is blown onto the item by a blast of air.

A is an incorrect answer because tamp-down is a label-placing technique in which the label is simply pressed against the item.

B is an incorrect answer because wipe-on is a label-placing technique in which the label is tamped down with the help of a foam roller.

D is an incorrect answer because direct thermal is a printing technique that produces a printed image using heat-sensitive paper when the paper passes over the thermal print head.

72. Answer:

A. 3

B. 4

C. 2

D. 1

73. Answer: C

C is the correct answer because there can be (and usually is) more than one root cause for a problem.

A, B, and **D** are incorrect answers because these are correct statements about root-cause analysis.

74. Answer: B

B is the correct answer because multiple tags on the same item are not necessarily going to cause tag failure.

A, C, and **D** are incorrect answers because these are valid reasons for a tag failure.

75. Answer: D

D is the correct answer because DHCP is a part of a TCP/IP network but not a defining part of the EPCglobal network.

A, B, and **C** are incorrect answers because these are elements of the EPCglobal network.

76. Answer: B

B is the correct answer because when you want to troubleshoot a device, before doing anything else it's always a good idea to restart it to see if the problem disappears.

A, C, and **D** are incorrect answers because one rule of troubleshooting is to try the simplest step first.

77. Answer: C

C is the correct answer because there is no such monitoring as root-cause monitoring; root-cause analysis is a technique to find the root cause of a problem.

A, B, and **D** are incorrect answers because all these are valid types of monitoring.

78. Answer: A

A is the correct answer because Simple Network Management Protocol (SNMP) is a TCP/IP protocol used to manage devices connected to TCP/IP networks.

B, C, and **D** are incorrect answers because these are false statements about SNMP.

79. Answer: D

D is the correct answer because low tag traffic volume is not the cause of read errors.

A, B, and **C** are incorrect answers because all these can be indicated by read errors. For example, low signal strength will provide shorter read range, and improper placement of tags will prevent the reader from reading the tags.

80. Answer: A

A is the correct answer because the change in read error rate indicates instability of the system.

B and **C** are incorrect answers because average tag traffic volume and read error rates by themselves are not any indication of the stability or instability of a system.

D is an incorrect answer because each RFID system runs at a predetermined operating frequency and does not measure the stability of the system.

81. Answer: C

C the correct answer because MTBF stands for *mean time between failures* and is not the cause of hardware failures.

A, B, and **D** are incorrect answers because electrostatic discharge (ESD), unregulated power supply, and harsh environmental conditions can damage a hardware device, which will cause hardware failure.

Bibliography and Resources

- *EPCglobal Glossary Version 6.0*, May 2005. Web Site: www.epcglobalinc.org
- Federal Standard 1037C: Glossary of Telecommunication Terms, 1996.
- IBM WebSphere RFID Solution, International Business Machines Corporation, 2004.
- NEMA 250: Enclosures for Electrical Equipment (1000 Volts Maximum), A standard by the National Electrical Manufacturers Association, 2003.
- *RFID for Dummies*, Wiley Publishing, Inc., 2005.
- RFID Smart Label Developer's Kit and Smart Label Pilot Printer Quick Setup Guide; Printronix.
- *RFID Journal*. Web Site: www.rfidjournal.com
- *RFID Essentials*, O'Reilly Media Inc., 2006.

Index

U

V

W

Z